W.D. Wallis

# Mathematics in the Real World

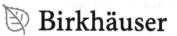 Birkhäuser

W.D. Wallis
Department of Mathematics
Southern Illinois University
Carbondale, IL, USA

ISBN 978-1-4939-4833-8          ISBN 978-1-4614-8529-2 (eBook)
DOI 10.1007/978-1-4614-8529-2
Springer New York Heidelberg Dordrecht London

Mathematics Subject Classification (2010): 05-01, 15-01, 60-01, 62-01, 91B12, 97-01

Printed on acid-free paper

Springer is part of Springer Science+Business Media (www.birkhauser-science.com)

*This book is dedicated to the memory of Rolf Rees and Tom Porter, two colleagues who left us far too soon.*

# Preface

In recent years, there has been a noticeable increase in the number of college courses designed to introduce mathematics to non-math students. I am not talking about the sort of mathematics studied at school, a glorified version of arithmetic with sole letters thrown in to represent numbers, but real mathematics—the study of patterns and structures (and yes, sometimes numbers are involved) that arise in our everyday lives.

This book is designed as a text for such a course, a one-semester course designed for students with a minimal background in mathematics.

## *Outline of Topics*

The first two chapters discuss the basics of numbers and set theory. Chapter 1 (*Numbers and Sets*) introduces the idea of a set and a subset (a part of a set); the sets of numbers that we use are discussed. Much of this material will be familiar to the reader, but you may see these topics differently after you see how the ideas are connected. We then introduce the idea of a base for numbers; we normally use base 10, but other bases are useful for dealing with computers. The rest of the chapter is about representing and using numerical data: summation, some notation for summations, and some ways of combining sets are introduced, Venn diagrams are discussed, and finally we look at arrays of numbers and define matrices.

Counting is covered in Chap. 2 (*Counting*). The concept of an event, a collection of possible outcomes, is introduced. Summations, which were a topic in Chap. 1, are studied further. We also distinguish between counting the number of selections or combinations—subsets of a given size—and counting arrangements or permutations—subsets where the order of the elements is important.

Counting leads naturally to probability theory. In Chap. 3 (*Probability*) we introduce the basic ideas of probability. Randomness and random events are defined. A number of examples involving dice and playing cards are included. In order to discuss compound events, tree diagrams (similar to family trees, which we'll mention later) are introduced. We go through two examples where probabilities are different from what many people expect—the birthday problem and the Monty Hall problem. Finally, the more mathematical concept of a probability model is introduced.

The next two chapters deal with statistics, and in particular those aspects (sampling, polls, predictions) that we often see in everyday life. As a first step toward describing statistics, Chap. 4 (*Data: Distributions*) examines the way in which numerical data about our world is collected and described. The ideas of a population and taking a sample from that population are introduced, together with ways of displaying this information—the dotplot, histogram, and boxplot—and parameters that describe the information, such as means, medians, and quartiles. Probability distributions are introduced, as a way of representing the probability model of a phenomenon; in particular, we look at the normal distribution.

Chapter 5 (*Sampling: Polls, Experiments*) looks at how statistics impinges on our world: estimating numerical properties of our society and deciding how reliable are those estimates. A sample is the group we examine to study a property (such as family income and height). The obvious question is: how reliable are our estimates in describing the whole population? We discuss the reliability of methods of sampling and ways of obtaining data, by observation or controlled experiments. Experimental designs are discussed briefly, and in particular Latin squares are studied. The relation of these designs to sudoku puzzles is mentioned.

We next look at graph theory. This is the study of (linear) graphs. A graph consists of a set of objects (called *vertices*) and a set of connections between pairs of vertices (called *edges*). For example, the vertices might represent cities and an edge joining two of them might mean there is a nonstop airline flight between those two cities; another example is the family tree, mentioned earlier, where vertices are people and edges represent parent–child connections.

In Chap. 6 (*Graphs: Traversing Roads*), we start by modelling roads as edges and look at the problem of traversing all roads in an area. Vertices might represent cities, parts of a town, or intersections. The Königsberg bridge problem asks whether one can traverse all roads in an area without any repeats; each road is to be traveled exactly once. If this is not possible, one can try to cover each road at least once, with the minimum possible number of repeats. Some basic ideas about graphs (adjacency, multiple edges, simplicity, connectedness) are discussed.

Chapter 7 (*Graphs: Visiting Vertices*) starts by defining some special types of graphs, such as paths, cycles, and bipartite graphs. The problem of visiting all towns in an area exactly once each—Hamilton's problem—and the problem of doing so as cheaply as possible—the traveling salesman problem—are introduced.

In Chap. 8 (*More About Graphs*) we discuss two further aspects of graph theory, trees and graph coloring. Trees are graphs reminiscent of the tree diagrams introduced in Chap. 3. In particular, spanning trees are defined and algorithms for finding minimum cost spanning trees are outlined. Coloring and chromatic number are introduced, as are some applications of graph coloring. The famous four-color theorem is discussed briefly. We use this chapter to illustrate a method of mathematical proof, the *proof by contradiction*, which some readers may choose to skip over on a first reading.

Chapters 9–11 deal with numbers we actually use in everyday life: credit card numbers, PINs, and so on. We also look at encoding and decoding, both for transmission of data and for secrecy.

In Chap. 9 (*Identification Numbers*) we look at the numbers we all use nowadays—account numbers, social security numbers, etc. We look at how these numbers are made up, and how they are used. In particular, the formulas for drivers' license numbers in Illinois and Florida are explained, as well as the check digits used in credit cards, postal money orders, and book identification numbers (ISBNs) to make sure the numbers are legitimate and have been transmitted correctly.

In our electronic world, much data are transmitted electronically. Computers basically transmit strings of digits, so it is necessary to encode and decode messages. Moreover, errors can occur, so check digits are required, just as they are for identification numbers. Chapter 10 (*Transmitting Data*) deals with this topic. One typical method, Venn diagram encoding/decoding, is examined in detail. This is an example of nearest neighbor decoding and is an example of a family of codes called Hamming codes. Variable-length codes, including Morse code and the genetic code, are also introduced. A surprising application of Hamming codes, the hat game, is described.

In Chap. 11 (*Cryptography*) we explore another reason for encoding material: secrecy. The history of secret writing, including the scytale and the Caesar cipher, is outlined. More modern techniques include the Vigenère method and substitution ciphers. Modular arithmetic is defined, and the RSA scheme of cryptography is studied.

The next two chapters are devoted to voting. In Chap. 12 (*Voting Systems*) some simpler voting systems and methods of deciding elections are discussed, starting with majority and plurality systems, then sequential voting and runoff elections. Preference profiles are defined. The Hare method for simple elections is described, together with the generalizations of the Hare method called instant runoff elections. Condorcet winners are defined, along with Condorcet's method of dealing with the case when there is no Condorcet winner. Sequential pairwise elections and pointscore methods are outlined.

Then in Chap. 13 (*More on Voting*) we describe two methods for elections when more than one candidate is to be elected: the generalized Hare method and approval voting. Then two methods of manipulating the vote are discussed. First is strategic voting, where voters might vote for their second favorite candidate to ensure that their least favorite candidate is not elected (called an insincere ballot); second is the introduction of amendments to change the final outcome. We close with an example of how different methods, even though they are fair, may give different results.

We finish by discussing various aspects of finance and related topics. Most readers have some idea of the mathematics of finance, but will be surprised by what they do not know. Chapter 14 (*The Mathematics of Finance*) covers simple and compound interest, the mathematics of compounding, and defines the annual percentage rate (the one lenders tell you about) and annual percentage yield (the one you actually pay). Geometric growth is introduced.

Chapter 15 (*Investments: Loans*) studies the mathematics of investments (regular savings) and (compound interest) loans. These two are similar: when you borrow, it is as if the lender is making an investment. Your equity when you borrow to cover a purchase is discussed. Two special types of loan, the add-on loan and the discounted loan, are explained.

The last chapter, Chap. 16 (*Growth and Decay*), looks at the growth of human and animal populations and at radioactive decay. These two are closely related. Both involve the limiting case of compounding, where the process is continuous. Exponential functions, exponential growth, and natural logarithms are discussed. The change in the cost of living is covered, as an extension of exponential growth.

## *Problems, Exercises, and Further Reading*

A number of worked examples, called Sample Problems, are included in the body of each section. Many of these are accompanied by a practice exercise labeled "Your Turn," designed primarily to test the reader's comprehension of the ideas being discussed. It is recommended that students work all of these exercises; complete solutions are provided at the end of the book.

The book contains a large selection of exercises, collected at the end of the chapters. They should be enough for students to practice the concepts involved; most of the problems are quite easy. Answers are provided. There is also a set of multiple-choice questions in each chapter.

Books of this kind often contain historical sketches, biographies, and the like. In many years of teaching, I have observed that students hardly ever read these items. Moreover, students are quite capable of looking up this information for themselves. For example, Googling "Samuel Morse" or "Morse code" will give as much information on the gentleman and his inventions as you could possibly need. So I chose to opt for a concise text; the Internet will provide all needed background and the student will not need to carry a huge book around.

Carbondale, IL                                                      W.D. Wallis

# Acknowledgments

This book owes a great deal to many colleagues and mathematicians with whom I have taught or discussed this material—too many to mention. I am also grateful for the constant support and encouragement of the staff at Birkhäuser, particularly Katherine Ghezzi.

Acknowledgments

# Contents

# Chapter 1
# Numbers and Sets

Most students will already know some of the material in this chapter, but even they should review some basics (sets, the standard number systems, and so on) so that we all use the same notation. We shall also introduce Venn diagrams, which will be a useful tool in more than one place.

## 1.1  Sets

All of mathematics rests on the foundations of set theory and numbers. We'll start this chapter by reminding you of some basic definitions and notations and some further properties of numbers and sets.

A *set* is any collection of objects. Sets abound in our lives—most children have owned a train set (a collection of engines, cars, track pieces, and so on); you have sets of CDs, sets of colored pencils, and so on. You could talk about all your friends as a set, or all your clothes.

The objects in the collection are called the *members* or *elements* of the set. If $x$ is a member of a set $S$, we write $x \in S$, and $x \notin S$ means that $x$ is *not* a member of $S$. One way of defining a set is to list all the elements, usually between braces; thus if $S$ is the set consisting of the numbers 0, 1 and 3, we could write $S = \{0, 1, 3\}$.

Another method is to use the *membership law* of the set: for example, since the numbers 0, 1 and 3 are precisely the numbers that satisfy the equation $x^3 - 4x^2 + 3x = 0$, we could write the set $S$ as

$$S = \{x : x^3 - 4x^2 + 3x = 0\}$$

(which we read as "the set of all $x$ such that $x^3 - 4x^2 + 3x = 0$"). Often we use a vertical line instead of the colon in this expression, as in

$$S = \{x \mid x^3 - 4x^2 + 3x = 0\}.$$

W.D. Wallis, *Mathematics in the Real World*, DOI 10.1007/978-1-4614-8529-2_1,
© Springer Science+Business Media New York 2013

This form is sometimes called *set-builder notation*.

> **Sample Problem 1.1** *Write three different expressions for the set with elements* 1 *and* −1.
> **Solution.** Three possibilities are $\{1,-1\}, \{x : x^2 = 1\}$, and "the set of square roots of 1". There are others.
> **Your Turn.** Write three different expressions for the set with the three elements 1, 2 and 3.

The definition of a set does not allow for ordering of its elements, or for repetition of its elements. For example, $\{1,2,3\}, \{1,3,2\}$ and $\{1,2,3,1\}$ all represent the same set. To handle problems that involve ordering, we define a *sequence* to be an ordered set. Sequences are denoted by using parentheses (round brackets) instead of the braces that we use for sets; $(1,3,2)$ is the sequence with first element 1, second element 3 and third element 2, and is different from $(1,2,3)$. Sequences can contain repetitions, and $(1,2,1,3)$ is quite different from $(1,2,3)$.

## 1.2   The Numbers We Use

We shall assume you know the basics of numbers—addition, multiplication and so on. Whole numbers are also called *integers*. We refer to a number as a *multiple* of another number if it equals the product of that number with an integer; for example the numbers 6, 12, 18, 36 are some of the multiples of 6, as are −6, −12, −24 (negative multiples) and even 0 (0 = 6 times 0).

It is natural to think of the numbers we use in terms of sets. The set $\mathbb{N}$ of all positive integers or *natural numbers* is the first number system we encounter; the natural numbers are used to count things. If one adds zero, to account for the possibility of there being nothing to count, and negatives for subtraction, the result is the set $\mathbb{Z}$ of *integers*. We write

$$\mathbb{N} = \{1,2,3,\ldots\}$$
$$\mathbb{Z} = \{\ldots -3,-2,-1,0,1,2,3,\ldots\}$$

The use of a string of dots (an *ellipsis*) means that the set continues without end. Such sets are called *infinite* (as opposed to *finite* sets like $\{0,1,3\}$). We also write $\{1,2,\ldots,20\}$ to mean the set of all positive integers from 1 to 20. When no confusion arises, the ellipsis means "continue in the obvious way."

The set $\mathbb{Q}$ of rational numbers consists of the ratios $p/q$, where $p$ and $q$ are integers and $q \neq 0$. In other words,

$$\mathbb{Q} = \{p/q : p \in \mathbb{Z}, q \in \mathbb{Z}, q \neq 0\}.$$

Each rational number has infinitely many representations as a ratio. For example,

$$1/2 = 2/4 = 3/6 = \dots$$

An alternative definition is that $\mathbb{Q}$ is the set of all numbers with a repeating or terminating decimal expansion. Examples are

$$1/2 = 0.5$$
$$-12/5 = -2.4$$
$$3/7 = 0.428571428571\dots$$

In the last example, the sequence 428571 repeats forever, and we denote this by writing

$$3/7 = 0.\overline{428571}.$$

The denominator $q$ of a rational number cannot be zero. In fact, division by zero is never possible. Some people—even, unfortunately, some teachers—think this is a made-up rule, but it is not. In fact, it follows from the definition of division. When we write $x = p/q$, we mean "$x$ is the number which, when multiplied by $q$, gives $p$." So what would $x = 2/0$ mean? There is no number which, when multiplied by 0, gives 2. Whenever you multiply any number $x$ by 0, you get 0. You can never get 2. How about $x = 0/0$? There are suitable numbers $x$, in fact *every* number will give 0 when multiplied by 0, but we wanted a single answer. So "$x = 0/0$" tells us nothing about $x$; it doesn't specify any number.

The integers are all rational numbers, and in fact they are the rational numbers with numerator 1. For example, $5 = 5/1$.

The positive integers are called "natural," but there is no special name for the positive rational numbers. However, we have a notation for this set, $\mathbb{Q}^+$. In general, a superscript $^+$ denotes the set of all positive members of the set in question.

The final number system we shall use is the set $\mathbb{R}$ of real numbers, which consists of all numbers which are decimal expansions, all numbers which represent lengths. Not all real numbers are rational; one easy example is $\sqrt{2}$. In fact, if $n$ is any natural number other than a perfect square (that is, $n$ is not one of 1, 4, 9, 16, ...), then $\sqrt{n}$ is not rational. Another important number which is not rational is $\pi$, the ratio of the circumference of a circle to its diameter.

Remember that every natural number is an integer; every integer is a rational number; every rational number is a real number. Do not fall into the common error of thinking that "rational number" excludes the integers, and so on. There are special words for such things. Rational numbers which are not integers are called *proper fractions*, and real numbers which are not rational are called *irrationals*.

This is not the end of number systems. For example, the set $\mathbb{C}$ of complex numbers is derived from the real numbers by including square roots of negative numbers, plus all the sums and products of the numbers. However, we will not encounter them in this book.

## 1.3   Bases

In ordinary arithmetic we use ten digits or one-symbol numbers $\{0, 1, 2, 3, 4, 5, 6, 7, 8, 9\}$ to write all the possible numbers. The symbol for "ten" is 10, meaning "once ten plus zero times one." For example, 243 means "twice ten-squared plus four times ten plus three." In symbols, we could write

$$243 = 2 \times 100 + 4 \times 10 + 3 = 2 \times 10^2 + 4 \times 10^1 + 3 \times 10^0.$$

To write numbers less than 1, we write 1/10, or $10^{-1}$, as .1; $1/100 = 10^{-2} = .01$, and so on. Ten is called the *base*.

In general, suppose $a_0, a_1, a_2$ and $b_1, b_2, b_3$ are any digits. When we write the number $\ldots a_2 a_1 a_0.b_1 b_2 b_3 \ldots$ it means

$$\ldots + a_2 \times 10^2 + a_1 x 10^1 + a_0 \times 10^0 + b_1 \times 10^{-1} + b_2 \times 10^{-2} + b_3 \times 10^{-3} + \ldots$$

Most people believe that we use 10 as the base of our number notation because people have 10 fingers and thumbs on their hands. But there is no special mathematical reason for choosing base 10. Historically, base 60 was used first, by the Sumerians and Babylonians.

We shall consider one other base, the base 2, because it arises in computer applications. Numbers written in base 2 are called *binary numbers*. To write binary numbers, only the two digits 0 and 1 are necessary.

We shall denote the base by putting the number in parentheses and then putting the base as a subscript. In that notation, $(101.11)_2$ means $1 \times 2^2 + 0 \times 2^1 + 1 \times 2^0 + 1 \times 2^{-1} + 1 \times 2^{-2}$, or in regular (base 10) notation $4 + 0 + 1 + .5 + .25$, equaling 5.75. So we could say $(101.11)_2 = (5.75)_{10}$. But we will usually omit the parentheses and subscript when the numbers are written in base 10.

**Sample Problem 1.2**   *What is* $(10111)_2$ *in base 10?*

**Solution.**        $(10111)_2 = 1 \times 2^4 + 0 \times 2^3 + 1 \times 2^2 + 1 \times 2^1 + 1$

$$= 16 + 0 + 4 + 2 + 1$$

$$= 23$$

**Your Turn.** What is $(1100100)_2$ in base 10?

**Sample Problem 1.3** *What is* $(.101)_2$ *in base 10? What is* $(10111.101)_2$ *in base 10?*

**Solution.**
$$(.101)_2 = 1 \times 2^{-1} + 0 \times 2^{-2} + 1 \times 2^{-3}$$
$$= 1 \times .5 + 0 \times .25 + 1 \times .125$$
$$= .5 + .125 = .625$$

Using this and the previous sample problem,

$$(10111.101)_2 = (10111)_2 + (.101)_2$$
$$= 123 + .625 = 23.625$$

**Your Turn.** What is $(10.111)_2$ in base 10?

In order to convert from base 10 to base 2, use continued division until you reach quotient 1, and record the remainders. Start with the final quotient (the 1) and read the remainders upward.

**Sample Problem 1.4** *What is* 108 *in base 2?*

**Solution.**     $108/2 = 54$, remainder 0
$54/2 = 27$, remainder 0
$27/2 = 13$, remainder 1
$13/2 = \phantom{0}6$, remainder 1
$6/2 = \phantom{0}3$, remainder 0
$3/2 = \phantom{0}1$, remainder 1

So you follow the initial 1 with 1101100, and $108 = (1101100)_2$.

**Your Turn.**
What is 91 in base 2?

Conversion of non-integers from base 10 to base 2 can also be done, but is more difficult.

## 1.4   Sums: Sigma Notation

Suppose you want to write the sum of the first 14 positive integers. You could write

$$1 + 2 + 3 + 4 + 5 + 6 + 7 + 8 + 9 + 10 + 11 + 12 + 13 + 14$$

but, instead of this clumsy form, it is more usual to write $1 + 2 + \ldots + 14$, assuming that the reader will take the ellipsis, or three dots, to mean "continue in this fashion until you reach the last number shown" (and, more importantly, hoping it is clear that "in this fashion" means each number in the sum is obtained by adding 1 to the preceding number).

There is a standard mathematical notation for long sums, which uses the Greek capital letter *sigma*, or $\Sigma$. The above sum is written

$$\sum_{i=1}^{14} i$$

which means we take the sum of all the values $i = 1, i = 2, \ldots,$ up to $i = 14$. This is called *sigma notation*, and $i$ is called the *index*. In the same way,

$$\sum_{i=1}^{6} i^2 = 1^2 + 2^2 + 3^2 + 4^2 + 5^2 + 6^2;$$

the notation means "first evaluate the expression after the $\Sigma$ (that is, $i^2$) when $i = 1$, then when $i = 2, \ldots,$ then when $i = 6$, and then add the results." More generally, suppose $a_1, a_2, a_3$ and $a_4$ are any four numbers. (This use of a *subscript* on a letter, like the 1, 2, $\ldots$ on $a$, is common in mathematics—otherwise we would run out of symbols!) Then

$$\sum_{i=1}^{4} a_i = a_1 + a_2 + a_3 + a_4.$$

(When the sigma notation is used in the middle of a printed line, rather than in a display, it usually looks like $\sum_{i=1}^{14} i$, so that the subscript and superscript don't mess up the line spacing.)

When we write $\sum_{i=1}^{n} a_i$, you could say we are using $a_i$ to mean a "general" or "typical" member of $\{a_1, a_2, \ldots, a_n\}$. This sort of usage is very common. When a set of numbers $\{a_1, a_2, \ldots, a_n\}$ is being discussed, we say a property is true "for all $a_i$" when we mean it is true for each member of the set.

Usually the sigma notation is used with a formula involving the index $i$ for the term following $\Sigma$, as in the following examples. Notice that the range need not start at 1; we can write $\sum_{i=j}^{n}$ when $j$ and $n$ are any integers, provided $j < n$. We can also break the sum into two or more parts; for example,

$$\sum_{i=1}^{n} a_i = \sum_{i=1}^{j} a_i + \sum_{i=j+1}^{n} a_i.$$

**Sample Problem 1.5** *Write out the following as sums and evaluate them:*

$$\sum_{i=1}^{5} i^2; \quad \sum_{i=3}^{5} i(i+1).$$

**Solution.**

$$\sum_{i=1}^{5} i^2 = 1^2 + 2^2 + 3^2 + 4^2 + 5^2$$

$$= 1 + 4 + 9 + 16 + 25$$

$$= 55;$$

$$\sum_{i=3}^{5} i(i+1) = 3 \cdot 4 + 4 \cdot 5 + 5 \cdot 6$$

$$= 12 + 20 + 30$$

$$= 62.$$

**Your Turn.** Write out the following as sums and evaluate them:

$$\sum_{i=3}^{5} i(i-1); \quad \sum_{i=2}^{6} i.$$

When you are given a sum, there can be more than one way to represent it in sigma notation.

**Sample Problem 1.6** *Write the following in sigma notation:*

(i) $2 + 6 + 10 + 14 + 18$;      (ii) $1 + 16 + 81$.

**Solution.** In (i), each term is greater by 4 than the preceding one, so we try an expression involving $4i$ for term $i$. Two simple possibilities are $4i - 2$ and $4i + 2$, giving solutions $\sum_{i=1}^{5} (4i - 2)$ and $\sum_{i=0}^{4} (4i + 2)$.

In (ii), observe that $16 = 2^4$ and $81 = 3^4$, so the answer is $\sum_{i=1}^{3} i^4$.

**Your Turn.** Write the following in sigma notation:

(i) $1 + 3 + 5 + 7 + 9$;      (ii) $8 + 27 + 64 + 125$.

It is easy to see that the following properties of sums are true:

(1) If $c$ is any given number, then $\sum_{i=1}^{n} c = nc$

(2) If $(a_i)$ and $(b_i)$ are any two sequences of numbers, both of length $n$, then

$$\sum_{i=1}^{n} (a_i + b_i) = \left( \sum_{i=1}^{n} a_i \right) + \left( \sum_{i=1}^{n} b_i \right).$$

(3) If $c$ is any given number and $(a_i)$ is any sequence of $n$ numbers, then

$$\sum_{i=1}^{n} (ca_i) = c \cdot \left( \sum_{i=1}^{n} a_i \right).$$

## 1.5   More About Sets

We defined the notation $s \in S$ to mean "$s$ belongs to $S$" or "$s$ is an element of $S$." If $S$ and $T$ are two sets, we shall write $T \subseteq S$ to mean that every member of $T$ is also a member of $S$. In other words, "If $s$ is any element of $T$ then $s$ is a member of $S$," or

$$s \in T \Rightarrow s \in S,$$

where $\Rightarrow$ is shorthand for *implies*. When $T \subseteq S$ we say $T$ is *subset* of $S$. Sets $S$ and $T$ are equal, $S = T$, if and only if $S \subseteq T$ and $T \subseteq S$ are both true. If necessary, we can represent the situation where $T$ is a subset of $S$ but $S$ is not equal to $T$—there is at least one member of $S$ that is not a member of $T$—by writing $S \subset T$, and we call $T$ a *proper* subset of $S$.

Suppose $R \subseteq S$ and $S \subseteq T$ are both true. Any member of $R$ will also be a member of $S$, which means it is a member of $T$. So $R \subseteq T$. This sort of rule is called a *transitive law*.

It is important not to confuse the two symbols $\in$ and $\subseteq$, or their meanings:

**Sample Problem 1.7** *Suppose $S = \{0, 1\}$. Which of the following are true:*
  (i) $0 \in S$, $\{0\} \in S$, $0 \subset S$,
  (ii) $\{0\} \subset S$, $0 \subseteq S$, $\{0\} \subseteq S$, $S \in S$,
  (iii) $S \subset S$, $S \subseteq S$?
**Solution.**
  (i) 0 is a member of $S$, but $\{0\}$ and $S$ are not, so $0 \in S$ is true but $\{0\} \in S$, and $S \in S$ are false.
  (ii) As 0 is a member of $S$, $\{0\} \subset S$ and $\{0\} \subseteq S$ are true. But 0 is not a *set* of elements of $S$, so $0 \subset S$ and $0 \subseteq S$ are false.
  (iii) $S \subseteq S$ is true, but $S \subset S$ would imply $S \neq S$, so it is false.

Among the standard number sets, many subset relationships exist. Every natural number is an integer, every integer is a rational number, and every rational number is a real number, so $\mathbb{N} \subseteq \mathbb{Z}$, $\mathbb{Z} \subseteq \mathbb{Q}$, $\mathbb{Q} \subseteq \mathbb{R}$. We could write all these relationships down in one expression:

$$\mathbb{N} \subseteq \mathbb{Z} \subseteq \mathbb{Q} \subseteq \mathbb{R}.$$

In fact, we know that no two of these sets are equal, so we could write

$$\mathbb{N} \subset \mathbb{Z} \subset \mathbb{Q} \subset \mathbb{R}.$$

Given sets $S$ and $T$, we define two operations: the *union* of $S$ and $T$ is the set

$$S \cup T = \{x : x \in S \text{ or } x \in T \text{ (or both)}\};$$

the *intersection* of $S$ and $T$ is the set

$$S \cap T = \{x : x \in S \text{ and } x \in T\}.$$

As a kind of opposite to the union, the notation $S \backslash T$ denotes the set of all members of $S$ that are *not* in $T$.

There is also a special relationship between subsets and the other operations. If $S$ is any subset of $T$, then $S \cap T = S$ and $S \cup T = T$.

Suppose two sets, $S$ and $T$, have no common element. Then $S$ and $T$ are called *disjoint*. In that case, $S \cap T$ is a set with no elements! There is no problem with the concept of such a set. We shall define the *empty* set, also called the *null* set, to be a set that has no elements. This set is denoted $\emptyset$. The set $\emptyset$ is unique and is a subset of every other set. Then "$S$ and $T$ are disjoint" means $S \cap T = \emptyset$.

Given sets $S$ and $T$, the notation $S \backslash T$ is used for the set formed by deleting from $S$ all the members that are also in $T$. Clearly $S \backslash T$ is the same as $S \backslash (S \cap T)$. If $S$ and $T$ are disjoint, then $S \backslash T = S$, while $S \backslash S = S$.

Finally, we can combine two sets $S$ and $T$ to form a new set called the *Cartesian product* $S \times T$. This consists of all the ordered pairs with the first element a member of $S$ and the second a member of $T$. For example, If $S = \{1,3\}$ and $T = \{2,3,4\}$ then

$$S \times T = \{(1,2),(1,3),(1,4),(3,2),(3,3),(3,4)\}.$$

In the following example, remember that a *perfect square* means a number of the form $n^2$, where $n$ is an integer.

**Sample Problem 1.8**  *In each case, are the sets S and T disjoint? If not, what is their intersection?*
  (i)  *S is the set of perfect squares, $T = \mathbb{R} \backslash \mathbb{R}^+$.*
  (ii)  *S is the set of all multiples of 5, T is the set of all multiples of 7.*
**Solution.**
  (i)  They are not disjoint, because 0 is a perfect square ($0 = 0^2$); $S \cap T = \{0\}$.
  (ii)  They are not disjoint. $S \cap T$ is the set of all multiples of 35.

## 1.6  Venn Diagrams

It is common to illustrate sets and operations on sets by diagrams. A set $A$ is represented by a circle, and it is assumed that the elements of $A$ correspond to the points (or some of the points) inside the circle. Such an illustration is called a *Venn diagram*; diagram methods were first used by Leibniz and Euler, but George Venn used the diagrams extensively and he formalized and unified diagram methods.

Elements common to two sets are shown in both corresponding circles, and the intersection of two sets is represented by the intersection of the two circles. For example, suppose $A$ is the set $\{1,2,3,4\}$ and $B = \{3,4,5,6,7\}$, then $A$ and $B$ would be shown in a diagram as follows:

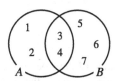

The intersection of the two circles contains the elements of $A \cap B = \{3,4\}$.

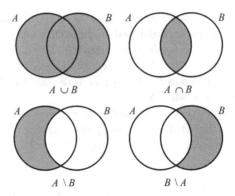

**Fig. 1.1**  Venn diagrams for combinations of sets

This same method can be used to represent more general sets. The diagrams in Fig. 1.1 show Venn diagrams representing $A \cup B$, $A \cap B$, $A \setminus B$ and $B \setminus A$; in each case, the set represented is shown by the shaded area. We do not know the individual members of the sets, but they could be represented by points in the diagram if they were known.

Venn diagrams are used to calculate the numbers of elements in sets. We write the numbers of elements in the areas of the diagram, instead of the elements themselves. Sometimes it is useful to represent objects that are not in any of the sets, so we usually show an outside area, represented as a rectangle, and write the number of other elements in there.

**Sample Problem 1.9**  *There are 60 students in Dr. Brown's Finite Mathematics course and 30 in his Calculus section. If these are his only classes, and if 20 of the students are taking both subjects, how many students does he have altogether? Represent the data in a Venn diagram.*

**Solution.** We use the notation $F$ for the set of students in the finite class and $C$ for Calculus. There are four areas in the Venn diagram: the set of students in both classes (the center area, $F \cap C$) has 20 members; the set of students in Finite Mathematics only (the left-hand enclosed area) has 40 members—subtract 20 from 60; the area corresponding to Calculus-only students has $30 - 20 = 10$ members; and the outside area has no members (only Dr. Brown's students are being considered). So Dr. Brown has 70 students; the diagram is

**Your Turn.** A survey shows that 12 newspaper readers buy the morning edition and seven buy the evening edition; three of these also bought the morning paper. Represent the data in a Venn diagram. How many of readers surveyed buy at least one edition?

This method can be used to find the number of elements in the set represented by any of the areas or unions of areas, provided there is enough information.

**Sample Problem 1.10**   120 *people were surveyed to find out whether newspaper advertisements or flyers were more efficacious in advertising supermarket specials.* 20 *of them said they pay no attention to either medium.* 50 *said they read the flyers, and* 15 *of those said they also check the newspapers. How many use the newspaper ads, in total?*

**Solution.** We do not know how many people read the newspaper advertisements but not the flyers. Suppose there are $X$ of them. Then we get the Venn diagram

(This time we have some members in the outside area.) In order for the total to add to 120, $X = 50$, so the total who use the newspaper ads is $X + 15 = 65$.

These methods can be applied to three or more sets.

**Sample Problem 1.11**   1,000 *people were asked about their morning vitamin intake. It was found that* 300 *take vitamin B,* 400 *take vitamin C,* 330 *take vitamin E,* 114 *take both B and C,* 250 *take both B and E,* 164 *take both C and E, and* 104 *take all three vitamins. How many take both B and E but do not take C? How many take none of these vitamins?*

**Solution.** We start with the diagram

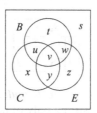

From the data, $v = 104$ and $u + v = 114$, so $u = 10$. Similarly, $w = 146$ and $y = 60$. Now $t + u + v + w = 300$, so $t = 40$. The other sizes are calculated similarly, and we get the diagram

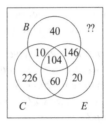

The cell corresponding to "*B* and *E* but not *C*" has 146 elements. There are 606 elements in total, so there are 394 people who take none.

**Your Turn.** In a survey of 150 moviegoers, it was found that 80 like horror movies, 75 like police procedurals, and 60 like romances. In total, 35 like both horror and procedurals, 25 like horror and romance, and 30 like procedurals and romance. There are 15 who like all three types of movie. Represent these data in a Venn diagram. How many like horror and police procedural movies, but do not like romances? How many like romances only? How many like none of these three types?

## 1.7   Arrays, Matrices

Suppose a movie theater sells three types of tickets—Adult (A), Student (S), and Child (C). The theater charges more after 6 PM, so tickets may also be classified as Day (D) or Evening (E). If 43 Adult, 33 Student and 18 Child tickets are sold for the afternoon session, and 78 Adult, 45 Student and 12 Child tickets are sold in the evening, the day's ticket sales could be represented by the following table:

|   | A | S | C |
|---|---|---|---|
| D | 43 | 33 | 18 |
| E | 78 | 45 | 12 |

An array of data like this is called a *rectangular array*, a *two-dimensional array* or simply an *array*. An array consisting only of numbers is called a *matrix*. In our example, the actual numbers are given by the matrix

$$M = \begin{vmatrix} 43 & 33 & 18 \\ 78 & 45 & 12 \end{vmatrix}.$$

We shall usually denote matrices by single upper-case letters. In general, matrices can be used whenever the data is classified in two ways; in the example, the classifications are ticket types (A, S, C) and session times (D, E). The horizontal layers are called *rows* and the vertical ones *columns* for example, the first row in the above matrix *M* is

$$\boxed{43\ 33\ 18}$$

and the second column is

$$\boxed{\begin{array}{c} 33 \\ 45 \end{array}}$$

If a matrix has $p$ rows and $q$ columns, we refer to it as a "$p$ by $q$ matrix" or say "the matrix is $p \times q$." The numbers $p$ and $q$ are the *dimensions* of the matrix.

**Sample Problem 1.12** *A furniture manufacturer makes tables and chairs. In January he made 200 tables and 850 chairs; in February, 300 tables and 1,440 chairs; in March, 140 tables and 880 chairs. Represent these data in an array.*
**Solution.** Write T for tables, C for chairs.

|     | T   | C    |
|-----|-----|------|
| Jan | 200 | 850  |
| Feb | 300 | 1440 |
| Mar | 140 | 880  |

**Your Turn.** In April, Joe's Autos sold 32 sedans and 16 pickups. In May, they sold 44 sedans and 12 pickups. Represent the two months' sales in an array.

It is possible that a matrix can have only one row or only one column. A matrix with one of its dimensions equal to 1 is called a *vector*. An $m \times 1$ matrix is a *column vector* of length $m$, while a $1 \times n$ matrix is a *row vector* of length $n$. The individual rows and columns of a matrix are vectors, which we call the *row vectors* and *column vectors* of the matrix.

The element in the $i$-th row and $j$-th column of a matrix is called the $(i, j)$ *entry* and it is denoted by the subscript $ij$; for example, a matrix $M$ has $(i, j)$ entry $m_{ij}$. Similarly, the $i$-th entry of a vector is denoted by subscript $i$.

It is possible for arrays to have more than two dimensions.

# Multiple Choice Questions 1

**1.** How many integers are there in the set $\{-3, 5/10, 0, 0.44, -4/2, 8.1, \sqrt{9}, \sqrt{27}\}$?

   A.  8                B.  6                C.  4                D.  2

**2.** How many rational numbers are there in the set in Question 1?

   A.  5                B.  7                C.  6                D.  8

**3.** Suppose $A = \{r, t, s\}, B = \{s, t, r, s\}, C = \{s, r, s, t\}$. Which of the sets are equal?

   A.  $A$ and $B$ only                B.  $B$ and $C$ only

   C.  $A$ and $C$ only                D.  all are equal

**4.** Write the sum $\displaystyle\sum_{i=1}^{2} \frac{8i}{i+17}$ without sigma notation and evaluate it.

   A.  $\dfrac{8}{1+17} + \dfrac{16}{2+17} = \dfrac{64}{171}$        B.  $\dfrac{8}{1+17} + \dfrac{8}{2+17} = \dfrac{148}{171}$

   C.  $\dfrac{8}{1+17} + \dfrac{16}{2+17} = \dfrac{220}{171}$       D.  $\dfrac{8}{1+17} + \dfrac{16}{2+17} = \dfrac{24}{37}$

**5.** If $\displaystyle\sum_{i=1}^{4} a_i = -5$ and $\displaystyle\sum_{i=1}^{4} b_i = 6$, what is $\displaystyle\sum_{i=1}^{4} a_i + b_i$?

   A.  1                B.  $-30$             C.  30             D.  $-1$

**6.** Evaluate $\displaystyle\sum_{i=2}^{5} 4i - 2$.

   A.  26               B.  42              C.  36             D.  48

**7.** If $\displaystyle\sum_{i=1}^{3} b_i = 5$, what is $\displaystyle\sum_{i=1}^{3} 2b_i$?

   A.  10              B.  30              C.  40             D.  11

**8.** There are eight students in the chess club and 12 students in the table tennis club. Exactly three students are in both. What is the total number of students in the two clubs?

   A.  20              B.  15              C.  17             D.  14

**9.** If $A$ and $B$ are two sets, then $A \cap B$ represents:

   A.  all elements in either $A$ and $B$

   B.  all elements in both $A$ and $B$

   C.  all elements that are in $A$ but not $B$

   D.  all sets that include both $A$ and $B$

**10.** Consider the two statements:

   **1.** $\{1,2\} \subseteq \{1,2,3\}$,

   **2.** $\{1,2\} \in \{1,2,3\}$

   Which of the following are true?

   A.  **1** but not **2**                B.  **2** but not **1**

   C.  both **1** and **2**              D.  neither **1** nor **2**

## Exercises 1

1. In each case, write the list of all members of the set.
   (i) $\{x : x$ is an even positive integer less than $12\}$
   (ii) $\{x : x$ is an odd integer between $-6$ and $6\}$
   (iii) $\{x : x$ is a color on the American flag $\}$
   (iv) $\{x : x$ is a month whose name starts with J$\}$
   (v) $\{x : x$ is a letter in the word "MISSISSIPPI"$\}$
2. Give three different expressions for the set with members 0, 1 and 2.
3. Which of the following statements are true?
   (i) $3 \in \{2,3,4,6\}$ $\qquad$ (ii) $4 \notin \{1,3,4,6\}$
   (iii) $5 \in \{2,3,4,6\}$ $\qquad$ (iv) $4 \in \{1,3,4,6\}$
4. Which of the following statements are true?
   (i) all natural numbers are integers;
   (ii) all integers are natural numbers;
   (iii) all rational numbers are real;
   (iv) all real numbers are rational.
5. For each of the following numbers, to which of the sets $\mathbb{N}, \mathbb{Z}, \mathbb{Q}, \mathbb{R}$ does it belong?
   (i) $1.308$ $\qquad$ (ii) $\sqrt{7}$ $\qquad$ (iii) $-7$
   (iv) $3.\overline{312}$ $\qquad$ (v) $\sqrt{-7}$ $\qquad$ (vi) $2 - 5/3$
6. For each of the following numbers, to which of the sets $\mathbb{N}, \mathbb{Z}, \mathbb{Q}, \mathbb{R}$ does it belong?
   (i) $-2.13$ $\qquad$ (ii) $\sqrt{5}$ $\qquad$ (iii) $\sqrt{4}$ $\qquad$ (iv) $2\pi$
   (iv) $1.8\overline{34}$ $\qquad$ (v) $2 + \sqrt{2}$ $\qquad$ (vi) $0$ $\qquad$ (vii) $\sqrt{2 + \sqrt{2}}$
7. Express each of the following numbers in base 10.
   (i) $(110111)_2$ $\qquad$ (ii) $(101010)_2$ $\qquad$ (iii) $(10001000)_2$
   (iv) $(.01)_2$ $\qquad$ (v) $(10001000.01)_2$ $\qquad$ (vi) $(110011.01)_2$
8. Express each of the following numbers in base 10.
   (i) $(101110)_2$ $\qquad$ (ii) $(101100)_2$ $\qquad$ (iii) $(1010100)_2$
   (iv) $(.101)_2$ $\qquad$ (v) $(101100.101)_2$ $\qquad$ (vi) $(1010100.101)_2$
9. Express each of the following numbers in base 2.
   (i) $35$ $\qquad$ (ii) $19$ $\qquad$ (iii) $201$ $\qquad$ (iv) $117$
10. Express each of the following numbers in base 2.
   (i) $22$ $\qquad$ (ii) $43$ $\qquad$ (iii) $184$ $\qquad$ (iv) $123$
11. Write each of the following as sums and evaluate them.
   (i) $\sum_{i=1}^{6} i^2 + 1$ $\qquad$ (ii) $\sum_{i=1}^{3} \frac{1}{10^i}$ $\qquad$ (iii) $\sum_{i=3}^{9} i(i-3)$
   (iv) $\sum_{i=2}^{5} \frac{1}{i}$ $\qquad$ (v) $\sum_{i=2}^{4} (-2)^i$ $\qquad$ (vi) $\sum_{i=1}^{4} 1 + (-1)^i$
   (vii) $\sum_{i=2}^{5} \frac{1}{i-1}$ $\qquad$ (viii) $\sum_{i=1}^{6} 1 + 3i$

**12.** Write each of the following as sums and evaluate them.

(i) $\displaystyle\sum_{i=1}^{4} i^2$      (ii) $\displaystyle\sum_{i=1}^{5} 1+(-1)^i$      (iii) $\displaystyle\sum_{i=3}^{6} 1-i$

(iv) $\displaystyle\sum_{i=6}^{15} 4$      (v) $\displaystyle\sum_{i=3}^{6} i(i-1)$      (vi) $\displaystyle\sum_{i=2}^{4} \frac{1}{i^2}$

**13.** Write each of the following sums in sigma notation.

(i)    $1+4+7+10+13$       (ii)   $2+6+10$

(iii)   $1-4+7-10+13$      (iv)   $0-1+4-9+16$

(v)    $3+7+3+7+3+7$       (vi)   $0+3+8+15$

**14.** Write each of the following sums in sigma notation.

(i)    $1+4+9+16+25$       (ii)   $2+5+8+11$

(iii)   $-1+2-3+4$           (iv)   $6+5+4+3+2+1$

(v)    $1+3+9+27+81$       (vi)   $4-6+8-10+12-14$

(vii)   $2+4+10+28+82$     (viii)   $4+8+12+16+20$

(ix)   $-1-2-3-4-5-6$    (x)   $11+17+23$

**15.** Consider the sets    $A_1 = \{1,2,3,4\}$,         $A_2 = \{2,4,6,8\}$,

                      $A_3 = \{3,4,5,8,9\}$,        $A_4 = \{4,3,2,1\}$,

                      $A_5 = \{1,2,3,4,5,6,7,8\}$.

(i) For which $i$ and $j$, if any, is $A_i \subseteq A_j$? For which $i$ and $j$, if any, is $A_i = A_j$?

(ii) Write down the elements of $A_i \cap A_j$ for every case where $i < j$.

**16.** Consider the sets    $A_1 = \{1,2,4\}$,     $A_2 = \{1,2,3,4,5\}$,

                      $A_3 = \{1,3,4,6\}$,     $A_4 = \{4,2,1\}$,

(i) For which $i$ and $j$, if any, is $A_i \subseteq A_j$? For which $i$ and $j$, if any, is $A_i = A_j$?

(ii) Write down the elements of $A_i \cap A_j$ for every case where $i < j$.

**17.** Suppose

$$A = \{2,3,5,6,8,9\} \quad B = \{1,2,3,4,5\} \quad C = \{5,6,7,8,9\}.$$

Write down the elements of

(i)   $A \cap B$                  (ii)   $A \cup C$

(iii)   $A \cup B \cup C$          (iv)   $A \cup (B \cap C)$

**18.** Suppose

$$A = \{a,b,c,d,e\} \quad B = \{a,c,e,g,i\} \quad C = \{c,e,f,i,o\}.$$

Write down the elements of

(i)   $A \cup B$                  (ii)   $A \cap C$

(iii)   $A \cup B \cup C$          (iv)   $A \cup (B \cap C)$

**19.** Is the given statement true or false?

(i)   $\{3,2\} = \{2,3\}$           (ii)   $4 \subseteq \{2,3,4,6\}$

(iii)   $\{3,2\} \subseteq \{1,2,3\}$       (iv)   $\{3,2\} \subseteq \{2,3\}$

20. Is the given statement true or false?
    (i) $2 \in \mathbb{Z}$  (ii) $\{1,2,3,4,5\} = \{5,4,3,2,1\}$
    (iii) $\mathbb{R} \notin \mathbb{Z}$  (iv) $\{4\} \in \{2,4,6\}$
21. Among 1,000 personal computer users it was found that 375 have a scanner and 450 have a DVD player attached to their computer. Moreover 150 had both devices.
    (i) How many had either a scanner or a DVD player?
    (ii) How many had neither device?
22. A survey was carried out. It was found that ten of the people surveyed were drinkers, while five were smokers; only two both drank and smoked, but seven did neither. How many people were interviewed?
23. Fifty people were asked about their book purchases. Twenty said they had bought at least one fiction book in the last week, 30 had bought at least one non-fiction book, and 10 had bought no books. Assuming that any book can be classified as either fiction or non-fiction, how many of the people interviewed had bought both fiction and non-fiction during the week?
24. Sixty people were asked about news magazines. It was found that 32 read *Newsweek* regularly, 25 read *Time* and 20 read *U.S. News and World Report*. Nine read both *Time* and *U.S. News and World Report*, 11 read both *Newsweek* and *Time*, and eight read both *Newsweek* and *U.S. News and World Report*. Eight of the people do not read any of the magazines
    (i) Represent the data in a Venn diagram. (Assume $x$ read all three.)
    (ii) How many read all three magazines?
    (iii) How many people read exactly one of the three magazines?
25. A psychologist has clients who receive individual attention (I) and others who are seen in group sessions (G). She has four private individual clients (P) and six who are sent to her by the court (C). Among her groups are 24 private and 12 court clients. Represent these data in an array.
26. A farmer needs to monitor the amounts of vitamins A, B and C in his chickens' diet. He buys two prepared food mixes. Each bag of food I contains 200 units of vitamin A, 100 units of vitamin B, and 250 units of vitamin C. Each bag of food II contains 250 units of vitamin A, 150 units of vitamin B, and 350 units of vitamin C. Represent this data in an array.
27. Give an example of a real number that is not a rational number.

# Chapter 2
# Counting

We often want to know the answer to the question, "How many?" In this chapter we shall look at some of the rules that help us to answer this question.

## 2.1 Some Counting Principles

One obvious question, given a set $S$, is "how many elements are there in $S$? We shall denote that number by $|S|$, the *order* of $S$.

We shall start with an example.

**Sample Problem 2.1** *Suppose there are 30 students in Dr. Green's finite mathematics course and 40 in his calculus section. If these are his only classes, and if 20 of the students are taking both subjects, how many students does he have altogether?*

**Solution.** We shall use the notation $S$ for the set of students in the finite mathematics class and $T$ for calculus. So $|S| = 30$ and $|T| = 40$, and $|S \cap T| = 20$. We want to know the number of elements of $|S \cup T|$.

If we wrote a list of all the students in the two classes, we would write $|S| + |T| = 30 + 40$ names. But we have duplicated the 20 names in $|S \cap T|$. Therefore the total number is

$$|S \cup T| = |S| + |T| - |S \cap T| = 30 + 40 - 20 = 50,$$

so he has 50 students.

This is the simplest case of a rule called the *principle of inclusion and exclusion*, and you can remember it in the following way. To list all members of the union of two sets, list all members of the first set and all members of the second set. This ensures that all members are included. However, some elements will be listed twice, so it is necessary to exclude the duplicates. If $S$ is the set of all the objects that

W.D. Wallis, *Mathematics in the Real World*, DOI 10.1007/978-1-4614-8529-2_2,
© Springer Science+Business Media New York 2013

have some property $A$ and $T$ is the set of all the objects with property $B$, then (2.1) expresses the way to count the objects that have *either* property $A$ *or* property $B$:

    (i)  count the objects with property $A$;

    (ii)  count the objects with property $B$;

    (iii)  count the objects with both properties;

    (iv)  subtract the third answer from the sum of the other two.

    Putting it another way, if you list all members of $S$, then list all the members of $T$, you will cover all members of $S \cup T$, but those in $S \cap T$ will be listed twice. To count all members of $S \cup T$, you could count all members of both lists, then subtract the number of duplicates. In other words,

$$|S \cup T| = |S| + |T| - |S \cap T|. \qquad (2.1)$$

    A similar argument can be applied to three or more sets. For example, suppose you want to know the number of elements in $S \cup T \cup W$. If you add $|S|$, $|T|$ and $|W|$, you have added all the elements of $S \cap T$ twice, and similarly those of $S \cap W$ and $T \cap W$. So subtract the orders of those sets. Then every element of one or two of the sets has been counted twice. But you are missing the elements of $S \cap T \cap W$—they were added three times and subtracted three times. So

$$|S \cup T \cup W| = |S| + |T| + |W| - |S \cap T| - |S \cap W| - |T \cap W| + |S \cap T \cap W|.$$

This can be generalized to any number of sets.

    A variation of this rule, called the *rule of sum*, can be simply expressed by saying "the number of objects with property $A$ equals the number that have both property $A$ and property $B$, plus the number that have property $A$ but not property $B$"; if we again define sets $S$ and $T$ to be the collections of all objects having properties $A$ and $B$ respectively, the rule is

$$|S| = |S \cap T| + |S \backslash T|. \qquad (2.2)$$

If we rewrite (2.2) as

$$|S \backslash T| = |S| - |S \cap T|$$

and substitute into (2.1), we obtain

$$|S \cup T| = |T| + |S \backslash T|. \qquad (2.3)$$

**Sample Problem 2.2** *Suppose the set $S$ has 35 elements, $T$ has 17 elements, and $S \cap T$ has seven elements. Find $|S \cup T|$ and $|S \backslash T|$.*

**Solution.** From (2.1) we see

$$|S \cup T| = |S| + |T| - |S \cap T| = 35 + 17 - 7 = 45.$$

From (2.2) we get

$$|S \backslash T| = |S| - |S \cap T| = 35 - 7 = 28.$$

**Your Turn.** Suppose $|S| = 40$, $|T| = 30$ and $|S \cap T| = 20$. Find $|S \cup T|$ and $|S \backslash T|$.

It is sometimes useful to break an event down into several parts, forming what we shall call a *compound event*. For example, suppose you are planning a trip from Los Angeles to Paris, with a stopover in New York. You have two options for the flight to New York: a direct flight with Delta or an American flight that stops in Chicago. For the second leg, you consider the direct flight with Air France, a British Airways flight through London, and Lufthansa stopping in Frankfurt. There are two ways to make the first flight and three to make the second, for a total of six combinations.

Suppose $S$ is the set of available flights for the first leg and $T$ is the set of available flights for the second leg. Then the flight combinations correspond to the members of the Cartesian product $S \times T$, and in this example

$$|S| = 2, \ |T| = 3, \ |S \times T| = |S| \times |T| = |2 \times 3| = 6.$$

The correspondence between compound events and Cartesian products applies in general. If $S$ is the set of cases where $A$ occurs, and $T$ is the set of cases where $B$ occurs, then the possible combinations correspond to the set $S \times T$, which has $|S| \times |T|$ elements. This idea is usually applied without mentioning the sets $S$ and $T$. Suppose the event $A$ can occur in $a$ ways and the event $B$ can occur in $b$ ways, then the combination of events $A$ and $B$ can occur in $ab$ ways. This very obvious principle is sometimes called the *multiplication principle* or *rule of product*. It can be extended to three or more sets.

**Sample Problem 2.3** *To open a bicycle lock you must know a three-number combination. You must first turn to the left until the first number is reached, then back to the right until the second number, then left to the third number. Any number from 1 to 36 can be used. How many combinations are possible?*
**Solution.** There are 36 ways to choose the first number, 36 ways to choose the second, and 36 ways to choose the third. So there are $36 \times 36 \times 36$ combinations.
**Your Turn.** Your debit card has a 4-digit PIN. If you can use any digits, how many PINs are possible?

**Sample Problem 2.4** *A true–false test consists of four questions. Assuming you answer all questions, how many ways are there to answer the test ?*
**Solution.** There are two ways to answer the first question, two ways to answer the second, two ways to answer the third, and two ways to answer the fourth. So the total is $2 \times 2 \times 2 \times 2 = 16$.

The multiplication principle only works when the events are performed independently—if the result of $A$ is somehow used to affect the performance of $B$, some combined results may be impossible. In the airline example, if the Delta flight leaves too late to connect with the Air France flight, then your choices are not independent, and only five combinations would be available.

Sometimes the order of the elements in a set is unimportant, but sometimes the order is significant. In that case we would like to know how many ways there are to order the elements of the set. For example, if you have three cards, an Ace, King and Queen (abbreviated as A, K, Q), they can be ordered in six ways: AKQ, AQK, KAQ, KQA, QAK, and QKA. The three-element set $\{A, K, Q\}$ has six orderings. So obviously any three-element set has three orderings.

More generally, suppose $S$ is a set with $n$ elements. How many different ways are there to order the elements of $S$?

We solve this by treating the ordering as a compound event with $n$ parts. There are $n$ ways to choose the first element of the ordered set. Whichever element is chosen, there remain $n - 1$ possible choices for the second element. When two elements have been selected, there are $n - 2$ choices for the third element.

In this way, we see that there are $n \times (n - 1) \times n - 2 \times \ldots \times 3 \times 2 \times 1$ ways to order $S$. This number is called $n$ factorial, and denoted $n!$. So

$$n! = n \cdot (n - 1) \cdot (n - 2) \cdot \ldots \cdot 3 \cdot 2 \cdot 1.$$

For convenience we define $0!$ to equal $1$.

The different ways of ordering the set $S$ are called *permutations of S*.

**Sample Problem 2.5** *Evaluate* $10!$.
**Solution.** $10! = 10 \times 9 \times 8 \times 7 \times 6 \times 5 \times 4 \times 3 \times 2 \times 1 = 362880$.
**Your Turn.** Evaluate $6!$.

**Sample Problem 2.6** *A committee of three people—chair, secretary, and treasurer—is to be elected by a club with* 11 *members. If every member is eligible to stand for each position, how many different committees are possible?*
**Solution.** We can treat the selection of the committee as a compound event with three parts: choose the chair, choose the secretary, and choose the treasurer. These parts can be performed in 11, 10, and 9 ways respectively. So there are $11 \times 10 \times 9$ committees possible.
**Your Turn.** What is the number of committees if there are 9 members?

This method is often combined with the multiplication principle.

**Sample Problem 2.7** *Four boys and four girls are to sit along a bench. The boys must sit together, as must the girls. How many ways can this be done?*
**Solution.** We treat this as a compound event with three parts. First, it is decided whether the boys are to be on the left or on the right. This can be done in two ways. Then the ordering of the boys is chosen. This can be done in $4! = 24$ ways. Finally, the girls are ordered. This can be accomplished in $4! = 24$ ways. So there are $2 \times 24 \times 24 = 1152$ arrangements.
**Your Turn.** How many ways could five boys and four girls be seated on two benches, if the boys must sit on the back bench and the girls on the front?

## 2.2  Arrangements

Suppose $S$ is a set with $s$ elements. We often need to know how many $k$-element *ordered* subsets or *k-sequences* or *arrangements* of size $k$ can be chosen from $S$. This number is denoted $P(s,k)$. For example, if $S$ is the three-element set $\{x,y,z\}$, the possible two-element ordered subsets are $xy, yx, xz, zx, yz$ and $zx$; so $P(3,2) = 6$.

In particular $P(s,s)$ denotes the number of $s$-sequences that can be chosen from an $s$-set, or the number of arrangements of the set $S$. But these are precisely the permutations of $S$. It follows that $P(s,s) = s!$. For this reason arrangements of size $k$ are often called *permutations of size k*.

Given an $s$-set $S = \{x_1, x_2, \ldots, x_s\}$, there are $s$ different sequences of length 1 on $S$, namely $(x_1)$, $(x_2)$, ..., and $(x_s)$. So $P(s,1) = s$. There are $s \times (s-1)$ sequences of length 2, because each sequence of length 1 can be extended to length 2 in $s-1$ different ways, and no two of these $s \times (s-1)$ extensions will ever be equal. So $P(s,2) = s(s-1)$. Similarly we find

$$P(s,3) = s \times (s-1) \times (s-2),$$

$$P(s,4) = s \times (s-1) \times (s-2) \times (s-3),$$

$$\ldots,$$

$$P(s,k) = s \times (s-1) \times (s-2) \ldots (s-k+1).$$

So $P(s,k)$ is calculated by multiplying, $s$, $s-1$, $s-2$, ... until there are $k$ factors.
It follows that

$$P(s,k) = s!/(s-k)!. \tag{2.4}$$

**Sample Problem 2.8**  *Calculate $P(10,3)$.*
**Solution.** There are two ways to calculate $P(10,3)$. We could say $P(10,3) = 10 \times 9 \times 8 = 720$. Or we could use the formula:

$$P(10,3) = 10!/7! = 362880/5040 = 720.$$

The first way is easier.
**Your Turn.** Calculate $P(6,4)$.

Several of the problems in the preceding section, such as those on selecting a committee, asked for the number of sequences of a certain length, and their solutions can sometimes be stated compactly by using arrangements.

**Sample Problem 2.9**  *A committee of three people—chair, secretary, and treasurer—is to be elected by a club with 14 members. If every member is eligible to stand for each position, how many different committees are possible?*
**Solution.** We can treat the committee as an ordered set of three elements chosen from the 14-element set of members. So the answer is $P(14,3)$, or $2,184$.

**Your Turn.** What is the number of committees if there are 12 members?

Sometimes an added condition makes the solution of a problem easier, not harder. For example, arranging people around a circular table is no more difficult than arranging them in a line, and sometimes easier.

Suppose $n$ people are to sit around a circular table. We start by arbitrarily labeling one seat at the table as "1," the one to its left as "2," and so on. Then there are $n!$ different ways of putting the $n$ people into the $n$ seats. However, we have counted two arrangements as different if one is obtained from the other by shifting every person one place to the left, because these two arrangements put different people in "1"; but they are clearly the same arrangement for the purposes of the question. Each arrangement is one of a set of $n$, all obtained from the others by shifting in a circular fashion. So the number of truly different arrangements is $n!/n$, which equals $(n-1)!$.

**Sample Problem 2.10** *How many ways can you make a necklace by threading together seven different beads?*
**Solution.** Suppose you put the beads on a table before threading them. There would be $(n-1)! = 6! = 720$ ways to arrange them in a circle. However, after the beads are threaded, the necklace could be flipped over, so every necklace has been counted twice (for example, $abcdefg$ and $agfedcb$ are the same necklace). Therefore, the total number is $6!/2 = 360$.
**Your Turn.** How many ways could the three boys and four girls be arranged around a circular table if the boys must sit together and the girls as well?

**Sample Problem 2.11** *Colette's Copying Company has eight photocopying machines and seven employees who can operate them. There are four copying jobs to be done simultaneously. How many ways are there to allocate these jobs to operators and machines?*
**Solution.** Call the jobs $A, B, C, D$. Choose two arrangements: first, which four operators should do the jobs; second, which machines should be used. The operator choice can be made in $P(7,4)$ ways, and the machines in $P(8,4)$ ways. In each case, the first member of the sequence is the one allocated to job $A$, the second to job $B$, and so on. There are $P(7,4) \times P(8,4) = 7 \times 6 \times 5 \times 4 \times 8 \times 7 \times 6 \times 5 = 1,411,200$ ways.

**Sample Problem 2.12** *The club in Sample Problem 2.9 wishes to elect a by-laws committee with three members—Chair, Secretary, and Legal Officer—and requires that no members of the main club committee be members of the by-laws committee. In how many different ways can the two committees be chosen?*
**Solution.** Suppose the main committee is chosen first. There are $P(14,3) = 2,184$ ways to do this. After the election, there are 11 members eligible for election to the by-laws committee, so it can be chosen in $P(11,3) = 990$ ways. So there are a grand total of $2,184 \times 990 = 2,162,160$.

In the preceding Sample Problem we see that, even in small problems, the numbers get quite large. It might be better to report the answer in its factored form, as $14 \times 13 \times 12 \times 11 \times 10 \times 9$. This form of answer also makes it clear that we could have solved the problem by treating the two committees as one six-member sequence, with $P(14,6)$ possible solutions.

In some cases, we want to talk about collections of objects with repetitions. For example, consider the letters in the word *ASSESS*. In how many ways can you order these letters? The set of letters involved is $\{A,S,E\}$, but there are six letters in the word, and the orderings will have six elements. For clarity, we often talk of *distinguishable* and *indistinguishable* elements. We could say the set of letters in *ASSESS* has six elements: four (indistinguishable) $S$'s, an $A$, and an $E$.

One way to tackle these problems is to assume the "indistinguishable" objects can be distinguished, and then take this into account. For example, if the letters in the word *ASSESS* were written on tiles with numbers as subscripts, like scrabble tiles, we could label them so that no two copies of the same letter get the same subscript, for example $A_1 S_1 S_2 E_1 S_3 S_4$. Then all the six letters are different, and there are 6! orderings. Say you have each of these orderings written on slips of paper. Now collect together into one pile all the slips that differ only in their subscripts. For example, $A_1 E_1 S_1 S_2 S_3 S_4$ and $A_1 E_1 S_2 S_1 S_3 S_4$ will be in the same pile, as will $A_1 E_1 S_2 S_3 S_1 S_4$, $A_1 E_1 S_4 S_2 S_3 S_1$, and several others. In fact, we can work out how many slips there are in a pile. There are four letters $S$, and one each of the others. Two slips will be in the same pile when they have the letters in the same order, but the subscripts on the $S$'s are in different order. There are $4! = 24$ ways to order the four subscripts, so there are 4! slips in each pile. Therefore, there are $6!/4! = 30$ piles. Two orderings can be distinguished if and only if their slips are in different piles, so there are $6!/4! = 30$ distinguishable orderings of *ASSESS*.

The same principle can be applied with several repeated letters. For example, if *SUCCESS* is written as $S_1 U_1 C_1 C_2 E_1 S_2 S_3$, we see that there are 2! ways of ordering the $C$'s and 3! ways if ordering the $S$'s, so each pile will contain $3! \times 2!$ slips, and the number of distinguishable orderings is $7!/(3! \times 2!) = 420$.

**Sample Problem 2.13** *In how many distinguishable ways can you order the letters of the word MISSISSIPPI?*
**Solution.** There are one $M$, four $I$'s, four $S$'s and two $P$'s, for a total of 11 letters. So the number of orderings is $11!/(4! \times 4! \times 2!)$ or 34650.
**Your Turn.** In how many distinguishable ways can you order the letters of the word *BANANA*?

## 2.3 Selections

Suppose you are giving a party, and you want to order three different pizzas from a list of 12 types that your local store sells. How many ways can you make your choice? If you were to list the three types, starting with your first choice, then the second, and finally the third, there would be 12 possible first choices, 11 second (you want different types, so no repeats are allowed), and 10 third. So the number of ways is $12 \times 11 \times 10 = 1320$. Essentially, you are calculating $P(12,3)$. But there would be six possible lists that give the same set of three pizzas, in different order. So there are really $1320/6 = 220$ possible choices.

Essentially, you are calculating the number of possible sets of three types of pizza you could choose from a set of 12 types. There are a number of situations similar to this: given a set $S$, we want to know how many different subsets of a given size are contained in $S$. These are called *selections* or *combinations*. We shall write $C(s,k)$ or $\binom{s}{k}$ for the number of different $k$-subsets of an $s$-set; it is usual to read the symbol as "$s$ choose $k$." $\binom{s}{k}$ is often called the *choice function* (of $s$ and $k$).

We can use the formula (2.4) to derive expressions for the numbers $C(s,k)$. Suppose $S$ is a set with $s$ elements. It is clear that every $k$-set that we choose from $S$ gives rise to exactly $k!$ distinct $k$-sequences on $S$ and that the same $k$-sequence never arises from different $k$-sets. So the number of $k$-sequences on $S$ is $k!$ times the number of $k$-sets on S, or

$$\binom{s}{k} = \frac{P(s,k)}{k!} = \frac{s!}{(s-k)!k!} \tag{2.5}$$

When calculating $\binom{s}{k}$ in practice, you would usually calculate $P(s,k)$, then divide by $k!$. So

$$\binom{s}{k} = \frac{s \times (s-1) \times (s-2) \times \ldots \times (s-k+1)}{1 \times 2 \times 3 \times \ldots \times k}.$$

There are $k$ factors in the denominator and in the numerator.

We agreed to say $0! = 1$. In combination with (2.5) this yields $\binom{s}{0} = 1$. This makes sense: it is possible to choose *no* elements from a set, but one cannot imagine different ways of doing so. We also define $\binom{s}{k} = 0$ if $k > s$. Again this makes sense—there is no way to choose more than $s$ elements from an $s$-set.

**Sample Problem 2.14** *Calculate $C(8,5)$ and $\binom{6}{6}$.*
**Solution.**

$$C(8,5) = \frac{8 \times 7 \times 6 \times 5 \times 4}{5 \times 4 \times 3 \times 2 \times 1} = 56.$$

$$\binom{6}{6} = \frac{6!}{0! \times 6!} = 1.$$

There is no need for calculation: the terms $6!$ in the numerator and denominator cancel.
**Your Turn.** Calculate $C(9,5)$ and $\binom{6}{0}$.

**Sample Problem 2.15** *A student must answer five of the eight questions on a test. How many different ways can she answer, assuming there is no restriction on her choice and the order in which she answers them is unimportant?*
**Solution.** $\binom{8}{5} = 56$ ways.
**Your Turn.** How many ways can she answer if she must choose five, one of which is Question 1?

**Sample Problem 2.16** *Computers read strings consisting of the digits 0 and 1. Such a string with k entries is called a k-bit string. How many 8-bit strings are there that contain exactly five 1s?*
**Solution.** To specify a string, it is sufficient to say which positions have 1s. There are $C(8,5)$ choices, so the answer is $C(8,5) = 56$.
**Your Turn.** How many 8-bit strings contain exactly four 1s?

**Sample Problem 2.17** *How many ways can a committee of three men and two women be chosen from six men and four women?*
**Solution.** The three men can be chosen in $\binom{6}{3}$ ways; the two women can be chosen in $\binom{4}{2}$ ways. Using the multiplication principle, the total number of committees possible with no restrictions is

$$\binom{6}{3} \times \binom{4}{2} = \frac{6!}{3!3!} \times \frac{4!}{2!2!}$$
$$= 120.$$

**Your Turn.** You wish to borrow two mystery books and three westerns from your friend. He owns five mysteries and seven westerns. How many different selections can you make?

**Sample Problem 2.18** *How many different "words" of five letters can you make from the letters of the word REPUBLICAN, if every word must contain two different vowels and three different consonants?*
**Solution.** The three consonants can be chosen in $\binom{6}{3} = 20$ ways, and the vowels in $\binom{4}{2} = 6$ ways. After the choice is made, the letters can be arranged in $5! = 120$ ways. So there are $20 \times 6 \times 120 = 14400$ "words."
**Your Turn.** What is the answer if you use the word *DEMOCRAT*?

## Multiple Choice Questions 2

1. If $|S| = 5$, $|T| = 7$ and $|S \cap T| = 4$, what is $S \cup T$?
   A.  2                  B.  6                  C.  4                  D.  8
2. If $|S| = 5$, $|T| = 7$ and $|S \cup T| = 10$, what is $S \cap T$?
   A.  2                  B.  6                  C.  4                  D.  8
3. Say there are 25 books on your shelf, each either hardback or paperback. If 10 are paperback, how many are hardback?
   A.  10                 B.  25                 C.  15                 D.  40
4. Judy wants one cereal, one juice and one coffee for breakfast. If she can choose from five cereals, three juices and two coffees (either caffeinated or decaf), how many different breakfasts are possible?
   A.  5                  B.  30                 C.  10                 D.  60
5. You have six books on the shelf at the back of your desk. In how many ways can they be ordered?
   A.  6                  B.  120                C.  720                D.  5040
6. Five people are to sit at a round table. How many ways can they be seated?
   A.  12                 B.  60                 C.  24                 D.  120
7. In how many ways can you arrange the letters of the word *LIFELINE*?
   A.  120                B.  5040               C.  6720               D.  40320
8. Eight students participate in a car wash for charity. In how many ways can you choose two of the students to hold the signs advertising the car wash?
   A.  8                  B.  16                 C.  28                 D.  56
9. There are nine students from whom four are going to be chosen to represent their school at a conference. If Jack, Anna or Chris, but only one of them, must be chosen, in how many ways can the students be chosen to go to the conference?
   A.  60                 C.  2160               B.  126                D.  3024
10. Evaluate $C(4,2) \times C(5,3)$.
   A.  720                C.  60                 B.  480                D.  16

## Exercises 2

1. Suppose $|S| = 19$, $|T| = 11$ and $|S \cap T| = 8$. Find $|S \cup T|$ and $|S \backslash T|$.
2. Suppose $|S| = 22$, $|T| = 12$ and $|S \cup T| = 28$. Find $|S \cap T|$ and $|S \backslash T|$.
3. Suppose $|T| = 37$, $|S \cap T| = 7$ and $|S \backslash T| = 14$. Find $|S|$ and $|S \cup T|$.
4. Suppose $|S| = 44$, $|T| = 18$ and $|S \cap T| = 12$. Find $|S \cup T|$ and $|S \backslash T|$.
5. In a survey of 1100 voters it was found that 275 will vote in favor of a $\frac{1}{2}\%$ increase in sales tax for Public Safety funding, 550 will vote in favor of a $\frac{1}{4}\%$ increase in income tax for school funding, and 200 of these will vote for both tax increases.
   (i) How many favor the school tax but not the Public Safety tax?
   (ii) How many will vote for neither option?

6. Among 1000 telephone subscribers it was found that 475 have answering machines and 250 call waiting. Moreover 150 had both options.
   (i) How many had either an answering machine or call waiting?
   (ii) How many had neither option?

7. Out of 400 people surveyed, 100 said they plan to buy a new house within the next three years, and 200 expected to buy a new car in that period. Of these, 50 planned to make both kinds of expenditure. Use (2.1) and the rule of sum to find out how many planned to buy a new house but not buy a new car, and how many planned to buy a new car but not buy a new house.

8. There are three roads from town $X$ to town $Y$, four roads from town $Y$ to town $Z$, and two roads from town $X$ to town $Z$.
   (i) How many routes are there from town $X$ to town $Z$ with a stopover in town $Y$?
   (ii) How many routes are there in total from town $X$ to town $Z$?
   (Assume that no road is traveled twice.)

9. List all different permutations of the set $\{A,B,C\}$.

10. The three boys and four girls in the choir are to sit on two benches. The boys must sit on the back bench and the girls on the front. How many different ways can they be seated?

11. A multiple-choice quiz contains eight questions. Each has three possible answers.
    (i) If you must answer every question, how many different answer sheets are possible?
    (ii) If you may either answer a question or leave a blank, how many different answer sheets are possible?

12. In how many different ways can you order the letters of the word "BREAK"?

13. Calculate:
    (i) $P(8,3)$      (ii) $P(4,4)$      (iii) $P(5,4)$      (iv) $P(9,2)$

14. Three men and four women sit in a row. How many different ways can they do it if:
    (i) the men must sit together;
    (ii) the women must sit together?

15. John likes to arrange his books. He has four western, seven mystery and six science fiction books, all different.
    (i) In how many ways can he arrange them on a shelf?
    (ii) In how many ways can he arrange them if all the books on the same subject must be grouped together?

16. Your PIN number consists of four digits. No repetitions are allowed, and 0 is not to be used.
    (i) How many PIN numbers are possible?
    (ii) How many PIN numbers are smaller than 4000?
    (iii) How many PIN numbers are even?
    (iv) How many PIN numbers contain no number greater than 6?

**17.** A football league consists of eight teams. Each team must play each other team twice: once as home team, once as visitors.
  (i) How many games must be played?
  (ii) If each pair plays only once (you don't care which is the home team), how many games must be played?

**18.** Seven stereo systems are to be arranged in a line against the wall of the appliance department.
  (i) How many ways can this be done?
  (ii) How many ways can they be arranged if the most expensive model must be in the middle?

**19.** In how many ways can you arrange the letters of the following words?
  (i)  *TODDLER*     (ii)  *OFFERED*     (iii) *BORROW*
  (iv)  *ARROWROOT*    (v)  *MOOSEWOOD*   (vi)  *APPLESEED*

**20.** There are ten speakers in a debate, five on each side. It is agreed that the first speaker must speak in favor of the proposition, followed by a speaker against it, then one in favor, then one against. The remaining six speakers may speak in any order. In how many different ways can the debate be scheduled?

**21.** List all the selections of size 3 that can be made from the set $\{A,B,C,D,E\}$.

**22.** Calculate the following quantities:
  (i)  $C(8,3)$      (ii)  $C(9,4)$      (iii)  $\binom{6}{3}$

  (iv)  $C(7,3)$     (v)  $C(7,7)$      (vi)  $\binom{8}{6}$

**23.** The Student Council consists of six juniors and 12 seniors. A committee of two juniors and three seniors is to be formed. How many ways can this be done?

**24.** A test has 12 questions, and you must answer nine of them.
  (i)  How many ways can you choose which questions to answer?
  (ii) In how many ways can you make your choice if you must include Question 1 or Question 2 (or maybe both)?

**25.** A state lottery requires you to choose five different numbers from $\{1,2,\ldots,49\}$. The order in which the numbers are chosen does not matter.
  (i)  How many possible choices are there?
  (ii) The state then draws six different numbers. You win if all five of your numbers are chosen. How many of the possible choices are winners?

**26.** A businessman wishes to pack three different ties for a business trip. If he has six ties available, how many different selections could be made?

**27.** There are 18 undergraduates and 14 graduates in the Math club. A committee of five members is to be selected. Calculate how many ways this can be done, if:
  (i)  There is no restriction.
  (ii) There must be exactly two graduates.
  (iii) There must be at least two graduates and two undergraduates.

**28.** A Euchre deck of cards contains 25 cards: 6 spades, 6 hearts, 6 diamonds, 6 clubs, and a joker. A five-card hand is dealt.
  (i)  How many different hands are possible?
  (ii) How many of those hands contain only spades?
  (iii) How many hands contain three spades and two clubs?

(iv) How many hands contain the joker?

(v) How many hands contain only spades and clubs, at least one of each?

**29.** Suppose the Senate contains 49 Democrats and 51 Republicans. A committee of six must be chosen.

(i) How many different committees can be chosen?

(ii) How many of these possible committees contain exactly three Democrats and three Republicans?

# Chapter 3
# Probability

In this chapter we consider the exact meaning of our everyday word "chance."

We talk about chance in various ways: "there is a good chance of rain today," "they have no chance of winning," "there is about one chance in three," and so on. In some cases the meaning is very vague, but sometimes there is a precise numerical meaning. We shall use the word "probability" to formalize those cases where "chance" has a precise meaning, and we shall assign a numerical value to probability.

## 3.1  Some Definitions

One way to think about the probability that an event will happen: suppose the same circumstances were to occur a great many times. In what fraction of cases would the event occur? This fraction is the *probability that the event occurs*. So probabilities will lie between 0 and 1; 0 represents impossibility, 1 represents absolute certainty. Often people express probabilities as percentages, rather than fractions. For example, consider the question: What is the chance it will rain tomorrow? We could ask, if the exact circumstances (current weather, time of year, worldwide wind patterns, and so on) were reproduced in a million cases, in what fraction would it rain the next day? And while we cannot actually make these circumstances occur, in practice we can try to get a good estimate using weather records and geographical/geophysical theory.

Many problems will involve ordinary dice, as used for example in games like Monopoly. These have six faces, with the numbers 1–6 on them. Dice can be biased, so that one face is more likely to show than another. If we roll an ordinary, unbiased die, what is the probability of rolling a 5? The six possibilities are equally likely, so the answer is $\frac{1}{6}$. If the die were biased, you might try rolling a few hundred times and keeping records.

W.D. Wallis, *Mathematics in the Real World*, DOI 10.1007/978-1-4614-8529-2_3,
© Springer Science+Business Media New York 2013

Another idea that is commonly used in probability problems is the deck of playing cards. A standard deck has 52 cards; the cards are divided into four suits: Diamonds and Hearts are red, Clubs and Spades are black. Each suit contains an Ace, King, Queen, Jack, 10, 9, 8, 7, 6, 5, 4, 3 and 2, and these are called the 13 *denominations*. (The Ace also doubles as a 1.) When a problem says "a card is dealt from a standard deck" it is assumed that all possible cards are equally likely, so the probability of any given card being dealt is $\frac{1}{52}$. There are four cards of each denomination, so the answer to "what is the probability of dealing a Queen," or any other fixed denomination, is $\frac{4}{52}$, or $\frac{1}{13}$.

We say an event is *random* if you can't predict its outcome for sure. This does *not* mean the chances of different outcomes are equal, although sometimes people use the word that way in everyday English. For example, if a die is painted black on 5 sides, white on one, then the chance of black is $\frac{5}{6}$ and the chance of white is $\frac{1}{6}$. This is random, although the two probabilities are not equal.

When we talk about the *outcomes* of a random phenomenon, we mean the distinct possible results; in other words, at most one of them can occur, and one must occur. The set of all possible outcomes is called the *sample space*. Each different outcome will have a probability. These probabilities follow the following rules:

1. One and only one of the outcomes will occur.
2. Outcome $X$ has a probability, $P(X)$, and $0 \leq P(X) \leq 1$.
3. The sum of the $P(X)$, for all outcomes $X$, is 1.

## 3.2  Events and Probabilities

Much of the language of probability comes from considering experiments. An *experiment* is defined to be any activity with well-defined, observable outcomes or results. For example, a coin flip has the outcomes "head" and "tail." The act of looking out the window to check on the weather has the possible results "it is sunny," "it is cloudy," "it is raining," "it is snowing," and so on. Both of these activities fit our definition of an "experiment."

The set of all possible outcomes of an experiment is the *sample space*. Each subset of the sample space is called an *event*. For our purposes, an event is simply a set of outcomes. For example, when we talk about the weather we might say there are four possible outcomes: *sunny*, *cloudy*, *rain*, or *snow*. So we could say the sample space is the set {*sunny*, *cloudy*, *rain*, *snow*}. The event "today is fine" consists of the outcomes *sunny* and *cloudy*. The event "It is not snowing" consists of *sunny*, *cloudy*, and *rain*. The different events can occur at the same time (for example, on a sunny day), so we say the events are not *mutually exclusive*.

When we roll a die, we are primarily interested in seeing which number is uppermost, so the sample space is the set $S = \{1,2,3,4,5,6\}$. The event "an odd number is rolled," for example, is the subset $\{1,3,5\}$. There are $2^6$ events (the number of subsets of the six-element set $S$).

Remember, probability 0 means the particular event *cannot* occur—impossibility. And probability 1 means the event *must* occur—certainty.

The probability of an event equals the sum of the probabilities of its outcomes.

**Sample Problem 3.1** *Consider a die that is weighted so that 1 is rolled one time in 5 and all other rolls are equally likely. What is the probability of rolling an even number?*

**Solution.** $P(1) = 0.2$, so the probabilities of the other five rolls must add to 0.8. They are equal, so $P(2) = P(3) = P(4) = P(5) = P(6) = 0.16$. So the probability of an even roll is $0.16 + 0.16 + 0.16 = 0.48$.

**Your Turn.** What is the probability of a roll of 4 or less with this die?

**Sample Problem 3.2** *Betty and John roll a die. If the result is a 1 or 2, John wins $2 from Betty; otherwise Betty wins $1 from John. What is the event "John wins"?*

**Solution.** $\{1, 2\}$.

**Your Turn.** In the above game, what is the event "Betty wins"?

The elementary counting principles from Chap. 2 will come in handy when computing probabilities. For example, if event 1 has $x$ possible outcomes, event 2 has $y$ possible outcomes, and the two events are not related, then the compound event "event 1 followed by event 2" has $xy$ possible outcomes. For example, "roll a die, look at score, roll again" has $6 \times 6 = 36$ outcomes. If an act has $n$ possible outcomes, then carrying it out $k$ times in succession, independently, has $n^k$ possible outcomes.

One convenient way to represent this sort of situation is a *tree diagram*

Consider the experiment where a coin is tossed; the possible outcomes are "heads" and "tails" ($H$ and $T$). The possible outcomes can be shown in a diagram like the one in Fig. 3.1a. A special point (called a *vertex*) is drawn to represent the start of the experiment, and lines are drawn from it to further vertices representing the outcomes, which are called the *first generation*. If the experiment has two or more stages, the second stage is drawn onto the outcome vertex of the first stage, and all these form the *second generation*, and so on. The result is called a *tree diagram*. Figure 3.1b shows the tree diagram for an experiment where a coin is tossed twice.

**Sample Problem 3.3** *A coin is flipped three times; each time the result is written down. What is the sample space? Draw a tree diagram for the experiment.*

**Solution.** The sample space is

$$\{HHH, HHT, HTH, HTT, THH, THT, TTH, TTT\}.$$

The tree diagram is shown in Fig. 3.2a.

The situation is more complicated where the outcome of the first event can influence the outcome of the second event, and so on. Again, tree diagrams are useful. We modify the previous example slightly.

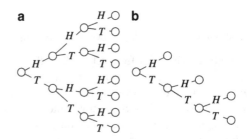

**Fig. 3.1** Tree diagrams

**Sample Problem 3.4** *A coin is flipped up to three times; you stop flipping as soon as a head is obtained. What is the sample space? Draw a tree diagram for the experiment.*

**Solution.** The sample space is $\{H, TH, TTH, TTT\}$. The tree diagram is shown in Fig. 3.2b.

**Fig. 3.2** Tree diagrams for Sample Problems 3.3 and 3.4

**Sample Problem 3.5** *An experiment consists of flipping three identical coins simultaneously and recording the results. What is the sample space?*

**Solution.** $\{HHH, HHT, HTT, TTT\}$. Notice that the order is not relevant here, so that for example, the events $HHT$, $HTH$ and $THH$ of Sample Problem 3.3 are all the same event in this case.

**Your Turn.** An experiment consists of flipping a quarter and noting the result, then flipping two pennies and noting the number of heads. What is the sample space? Draw a tree diagram.

One common application of tree diagrams is the *family tree*, which records the descendants of an individual. The start is the person, the first generation represents his or her children, the second the children's children, and so on. (This is why the word "generation" is used when talking about tree diagrams.)

Since events are sets, we can use the language of set theory in describing them. We define the union and intersection of two events to be the events whose sets are the union and intersection of the sets corresponding to the two events. For example, in Sample Problem 3.2, the event "either John wins or the roll is odd" is the set $\{1,2\} \cup \{1,3,5\} = \{1,2,3,5\}$. (For events, just as for sets, "or" carries the understood meaning, "or both".) The complement of an event is defined to have associated with it the complement of the original set in the sample space; the usual interpretation of the complement $\overline{E}$ of the event $E$ is "$E$ doesn't occur." Venn diagrams can represent events, just as they can represent sets.

The language of events is different from the language of sets in a few cases. If $S$ is the sample space, then the events $S$ and $\emptyset$ are called "certain" and "impossible." If $U$ and $V$ have empty intersection, they are disjoint sets, but we call them *mutually exclusive* events (to be consistent with ordinary English usage).

**Sample Problem 3.6** *A die is thrown. Represent the following events in a Venn diagram:*

A: *An odd number is thrown;*
B: *A number less than 5 is thrown;*
C: *A number divisible by 3 is thrown.*

**Solution.**

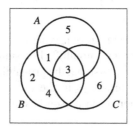

**Your Turn.** Repeat the above example for the events:

A: An even number is thrown;
B: John wins (in the game of Sample Problem 3.2);
C: Betty wins.

**Sample Problem 3.7** *A coin is tossed three times. Events A, B, and C are defined as follows:*

A: *The number of heads is even;*
B: *The number of tails is even;*
C: *The first toss comes down heads.*

*Write down the outcomes in A, B, C, $A \cup C$, $B \cap C$, and $\overline{C}$. Are any of the sets mutually exclusive? Write a brief description, in words, of $B \cap C$.*

**Solution.** $A = \{HHT, HTH, THH, TTT\}$, $B = \{HHH, HTT, THT, TTH\}$, $C = \{HHH, HHT, HTH, HTT\}$, $A \cup C = \{HHH, HHT, HTH, HTT, THH, TTT\}$, $B \cap C = \{HHH, HTT\}$, and $\overline{C} = \{THH, THT, TTH, TTT\}$. $A$ and $B$ are mutually exclusive, as are $A$ and $B \cap C$. $B \cap C$ consists of all outcomes with a head first and the other two results equal.

**Your Turn.** For the sets defined above, write down the outcomes in $A \cap C$, $B \cup C$, and $\overline{A}$.

Suppose all the different outcomes are equally likely. Say $S$ is the sample space, and $E$ is the set of all outcomes which we can describe as saying a certain event happens. For example, rolling a die, say the event is "score 5 or more." Then

$$S = \{1, 2, 3, 4, 5, 6\}$$
$$E = \{5, 6\}$$

and

$$P(\text{event happens}) = \frac{|E|}{|S|}.$$

(In the example, this equals 2/6.)

## 3.3  Two Interesting Examples

Probabilities can sometimes be surprising. We shall discuss two examples.

**Birthday Probabilities**

Suppose there are 30 people in a room. What is the probability that two of them share the same birthday (day and month)? (For simplicity, we shall ignore leap years.) Most people would say at first that the probability is small, but this is wrong.

First, list the 30 peoples' birthdays in the order of their names (alphabetical order). If there is no restriction, each person has 365 possible birthdays. So there are $365^{30}$ possibilities.

Suppose no two of them have the same birthday. There are 365 choices for the first person's birthday, 364 for the second, 363 for the third, and so on. With no repeats the total number of possible lists of birthdays is

$$365 \times 364 \times \ldots \times 336.$$

Therefore the probability of the event "no two have the same birthday" is

$$\frac{365 \times 364 \times \ldots \times 336}{365^{30}}$$

which is approximately .294. So the probability of the "birthday coincidence" is about

$$1 - .294 = .706,$$

or approximately 70%.

**The Game Show**

Our second example was made popular by columnist Marilyn Vos Savant.

Consider a TV game show: the contestant chooses between three doors marked doors 1, 2, and 3. Behind one door is a new car; behind each other is something almost worthless (a week's supply of Kleenex tissues; two movie tickets; a pet goat)—call it a booby prize. The contestant chooses a door, and she gets the prize behind it.

But wait! After the choice is announced but before door is opened, the host opens one of the other two doors (not the one the contestant chose). Behind the door we see a goat. The host then asks, "Do you want to stay with your original choice? Or would you rather switch to the third door?"

Well, should the contestant stay or switch? Or doesn't it matter? Most people would say it doesn't matter.

To analyze the problem, we need to agree on a few things.

1. On any given night the chance that the car is behind any particular door is one in three.
2. The host always opens a door to show a goat, then offers the switch.
3. If the contestant's first choice is the door with the car, there is an equal chance that the host will open either of the other two doors. He doesn't open the lower-numbered one more often, or anything like that.
4. The game is *always* played the same way (the host never skips the "open another door" part).

For ease of analysis, suppose the doors are numbered 1, 2, and 3. The contestant chooses door 1. We shall write $C1$ to mean "the car is behind door 1," $C2$ for "car behind door 2," and $C3$ for "car behind door 3." Similarly, $H2$ means "host opens door 2", and $H3$ means "host opens door 3". (He will not ever open door 1.)

Now suppose the game is played 600 times. We expect the car to behind each door in 200 cases. If the car is behind door 1, then we expect the host will open door 2 in 100 cases and open door 3 in 100 cases. In the 200 cases where the car is behind door 2, he always opens door 3; in the 200 cases where the car is behind door 3, he always opens door 2. So we can represent the data by

|      | $C1$ | $C2$ | $C3$ |
|------|------|------|------|
| $H2$ | 100  | 0    | 200  |
| $H3$ | 100  | 200  | 0    |

Now suppose the host opens door 2. This tells us that tonight is one of the 300 cases represented in the row H2. Of those 300 cases, the car is behind door 1 in 100 cases and behind door 3 in 200 cases. So the odds in favor of switching are 2 to 1. The same reasoning applies when he opens door 3. So it is always best to switch.

This problem is based on the game "Let's Make A = a Deal," with host Monty Hall, so it is often called the *Monty Hall problem*. Mr Hall has pointed out that in the real world, the conditions 1–4 do not always apply.

## 3.4 Probability Models

A *probability model* consists of a sample space together with the list of all probabilities of the different outcomes.

One example is a fair die, whose outcomes 1, 2, 3, 4, 5, 6 have probabilities

$$
\begin{array}{llll}
P(1) & = & \frac{1}{6} & \qquad\qquad P(2) & = & \frac{1}{6} \\
P(3) & = & \frac{1}{6} & \qquad\qquad P(4) & = & \frac{1}{6} \\
P(5) & = & \frac{1}{6} & \qquad\qquad P(6) & = & \frac{1}{6}
\end{array}
$$

This is a perfectly good probability model.

Another is a biased die, whose outcomes 1, 2, 3, 4, 5, 6 have probabilities

$$
\begin{array}{llll}
P(1) & = & 0.15 & \qquad\qquad P(2) & = & 0.19 \\
P(3) & = & 0.18 & \qquad\qquad P(4) & = & 0.16 \\
P(5) & = & 0.14 & \qquad\qquad P(6) & = & 0.18
\end{array}
$$

This is still a good probability model, because the probabilities still add to 1.

In some cases there is no numerical value associated with the outcomes—for example, predicting the weather. However, in many probability models, each outcome is a number, or has a numerical value. For example, this is true of rolling dice.

For example, suppose you are playing a game with three outcomes, call them $A, B, C$. If the outcome is $A$, you win \$10. If it is $B$, you win \$2. And if it is $C$, you lose \$15. The game is completely random, and observations show that $A$ and $C$ each occur 40% of the time, while $B$ is the result of 20% of plays.

If the game were played 100 times, your best guess would be that $A$ came up 40 times, $B$ 20 and $C$ 40. So someone who played 100 times might expect to win $\$(10 \times 40 + 2 \times 20)$, or \$440, and lose $\$15 \times 40 = \$600$. The net loss is \$160, or \$1.60 per play. You would say this is the *expected* cost of a play.

In general, suppose an occurrence has a numerical value associated with outcome, and a probability also. The *expected value* of the occurrence is found by multiplying each value by the associated probability, and adding. This is also called the mean value, or *mean* of the occurrence. The mean is most commonly denoted $m$.

**Sample Problem 3.8** *You are playing a game where you draw cards from a standard pack. If you draw a 2, 3, ..., or 10, you score 1 point; a Jack, Queen or King is worth 3; and an Ace is 5 points. How many points would you get for a typical draw?*

**Solution.** The probability of a draw worth 1 point is 36/52, or 9/13, since 36 of the 52 cards are worth 1 point, Similarly, the probabilities of 3 and 5 are 3/13 and 1/13 respectively. So the expected value is

$$
\frac{9}{13} \times 1 + \frac{3}{13} \times 3 + \frac{1}{13} \times 5 = 23/13.
$$

Similar calculations can be applied to any probability model where the outcomes have values associated with them.

# Multiple Choice Questions 3

1. Consider the statements
   X. An event can contain more than one outcome.
   Y. An outcome can occur in more than one event.
   Which of the following is true?
   A. X but not Y              B. Y but not X
   C. Both X and Y           D. Neither X nor Y

2. In an experiment, a die is rolled and a quarter is tossed at the same time. How many outcomes are possible?
   A. 24          B. 12          C. 6          D. 2

3. Exactly one of Terry, Chris, and Kim will attend a party. The probability Terry attends is 0.31 and the probability Chris attends is 0.5. What is the probability that Kim attends?
   A. 0.33          B. 0.5          C. 0.81          D. 0.19

4. A sample space consists of three outcomes, $X$, $Y$, and $Z$. Which of the following could be a legitimate assignment of probabilities to the outcomes?
   A. $P(X) = 0.3\ P(Y) = 0.6\ P(Z) = 0.9$
   B. $P(X) = 0.3\ P(Y) = 0.2\ P(Z) = 0.5$
   C. $P(X) = 0.3\ P(Y) = 0.3\ P(Z) = 0.3$
   D. $P(X) = 0.7\ P(Y) = 0.3\ P(Z) = 0$

5. Two fair dice are rolled and the sum rolled is recorded. What is the probability that the sum is 4?
   A. 1/3          B. 1/12          C. 4/11          D. 1/9

6. There are seven blue and six black socks in a drawer. One is pulled out at random. Find the probability that it is black.
   A. 6/13          B. 6/7          C. 1/2          D. 1/6

7. A fair coin is tossed three times. Find the probability of getting exactly two heads.
   A. 1/2          B. 1/3          C. 2/3          D. 3/8

8. A sample space has three outcomes, A, B, and C. The probability of outcome A is 0.39 and the probability of outcome B is 0.25. What is the probability of outcome C?
   A. 0.36          B. 0.33          C. 0.5          D. 0.64

9. In an experiment, a penny, a dime and a quarter are flipped simultaneously and the result (heads or tails) is recorded for each coin. How many members does the sample space contain?
   A. 8          B. 6          C. 2          D. 3

10. A game has three outcomes, $X$, $Y$, and $Z$, where $P(X) = 0.2$, $P(Y) = 0.3$, and $P(Z) = 0.5$. A player wins \$30 if $A$ occurs, wins \$40 if $B$ occurs, and loses \$20 if $C$ occurs. What is the expected value of one play of the game?
    A. \$8          B. −\$10          C. \$18          D. \$28

# Exercises 3

1. If the probability that it will rain today is 0.78, what is the probability that it will not rain?

2. Exactly one of three contestants will win a game show. The probability that John wins is 0.25 and the probability that Mary wins is 0.6. What is the probability that Shirley wins?

3. A student is taking a five-question "True"/"False" Test. If the student chooses answers at random, what is the probability of getting all questions correct?

4. Three (distinguishable) coins are flipped simultaneously and it is recorded whether each coin lands heads or tails. List the sample space.

5. A box contains five red, two blue, and three white marbles. One is selected at random. What is the probability that:
   (i)  The marble is blue;
   (ii)  The marble is not blue?

6. A single die is rolled. What is the probability that the number shown is:
   (i)  a 6?
   (ii)  smaller than 6?
   (iii)  6 or smaller?

7. Two fair dice are rolled. What are the probabilities of the following events?
   (i)  The total is 6;
   (ii)  The total is even;
   (iii)  Both numbers shown are even.

8. A fair coin is tossed four times. What are the probabilities that:
   (i)  At least three heads appear;
   (ii)  An even number of heads appear;
   (iii)  The first result is a head?

9. Two dice is rolled and the numbers showing are added. What is the probability that the sum is greater than 8?

10. An experimenter tosses four coins—two quarters and two nickels—and records the number of heads. For example, two heads on the quarters and one on the nickels is recorded $(2, 1)$.
    (i)  Write down all members of the sample space.
    (ii)  Write down all members of the following events:
         $E$ :  There are more heads on the quarters than on the nickels.
         $F$ :  There are exactly two heads in total.
         $G$ :  The number of heads is even.

11. An examination has two questions. Of 100 students 75 do Question 1 correctly and 70 do Question 2 correctly; 65 do both questions correctly.
    (i)  Represent these data in a Venn diagram.
    (ii)  A student's answer book is chosen at random. What is the probability that:
         (i)  Question 1 contains an error;
         (ii)  Exactly one question contains an error;
         (iii)  At least one question contains an error?

12. A bag contains two red, two yellow, and three blue balls. In an experiment, one ball is drawn from the bag and its color is noted, and then a second ball is drawn and its color noted.
    (i) Draw a tree diagram for this experiment.
    (ii) How many outcomes are there in the sample space?
    (iii) What are the outcomes in the event, "two balls of the same color are selected"?

13. A coin is tossed. If the first toss is a head, the experiment stops. Otherwise, the coin is tossed a second time.
    (i) Draw a tree diagram for this experiment.
    (ii) List the members of the sample space.
    (iii) Assume the coin is fair, that is there is a probability $\frac{1}{2}$ of tossing a head. What is the probability of the event, "the coin is tossed twice"?

14. A sample space contains four outcomes: $s_1, s_2, s_3, s_4$. In each case, do the probabilities shown form a probability model? If not, why not?
    (i)   $P(s_1) = .2, \ P(s_2) = .4, \ P(s_3) = .3, \ P(s_4) = .1$.
    (ii)  $P(s_1) = .2, \ P(s_2) = .4, \ P(s_3) = .6, \ P(s_4) = .2$.
    (iii) $P(s_1) = .1, \ P(s_2) = .2, \ P(s_3) = .5, \ P(s_4) = .2$.
    (iv)  $P(s_1) = .4, \ P(s_2) = .4, \ P(s_3) = .5, \ P(s_4) = -.3$.
    (v)   $P(s_1) = .1, \ P(s_2) = .2, \ P(s_3) = .3, \ P(s_4) = 1.2$.

15. Students in a course may receive grades A, B, C, D or F. Mary estimates the probabilities that she will receive the various grades as:

| Grade | A | B | C | D | F |
|-------|------|------|----|------|------|
| Prob | 0.01 | 0.05 | ?? | 0.35 | 0.04 |

What was her estimate of the probability of a C?

16. A die is rolled twice, and the results are recorded as an ordered pair.
    (i) How many outcomes are there in the sample space?
    (ii) Consider the events:
        E : The sum of the throws is 4;
        F : Both throws are even;
        G : The first throw was 3.
        (a) List the members of events E, F, and G.
        (b) Are any two of the events E, F, G mutually exclusive?
    (iii) Draw a tree diagram for this experiment.
    (iv) Represent the outcomes of this experiment in a Venn diagram. Show the events E, F, and G in the diagram.
    (v) Assuming the die is unbiased, what are the probabilities of E, F, and G?

17. There are 18 members in your club. There are eight men and ten women. The club committee has three members. One club member's name is selected at random. What is the probability that:
    (i) The person selected is on the committee;
    (ii) The person selected is male?

18. A regular deck of cards is shuffled and one card is dealt.
    (i) What is the probability that the card is red?
    (ii) What is the probability that the card is an Ace?

(iii)  What is the probability that the card is a red Ace?

(iv)  What is the probability that the card is either red or an Ace?

**19.** A card is drawn from a standard deck of cards, and then replaced. Then another card is drawn.

(i)  What is the probability that both are red?

(ii)  What is the probability that neither is red?

**20.** An experiment consists of studying families with three children. $B$ represents "boy", $G$ represents "girl" and for example $BGG$ will represent a family where the oldest child is a boy and the young children are both girls.

(i)  What is the sample space for this experiment?

(ii)  We define the following events:

      $E$ :  The oldest child is a boy;

      $F$ :  There are exactly two boys.

      (a)  What are the members of $E$ and $F$?

      (b)  Describe in words the event $E \cap F$.

(iii)  Draw a Venn diagram for this experiment, and show the events $E$ and $F$ on it.

**21.** Electronic components are being inspected. Initially one is selected from a batch and tested. If it fails, the batch is rejected. Otherwise a second is selected and tested. If the second fails, the batch is rejected; otherwise a third is selected. If all three pass the test, the batch is accepted.

(i)  Draw a tree diagram for this experiment.

(ii)  How many outcomes are there in the sample space?

(iii)  What is the event, "fewer than three components are tested"?

**22.** At a local grocery store, the number of people in checkout lines varies. The probability model for the number of people in a randomly chosen line is

| Number in line | 0 | 1 | 2 | 3 | 4 | 5 |
|---|---|---|---|---|---|---|
| Probability | 0.08 | 0.15 | 0.20 | 0.22 | 0.15 | 0.20 |

What is the mean number of people in a line?

**23.** Below is a probability model for the number of automobiles owned by a randomly chosen family in a large town. What is the mean number of automobiles owned?

| Number of cars | 0 | 1 | 2 | 3 |
|---|---|---|---|---|
| Probability | 0.05 | 0.55 | 0.30 | 0.10 |

What is the mean number of people in a line?

# Chapter 4
# Data: Distributions

Often we want to collect data about all members of some group—all salary earners, all cattle in a herd, all light bulbs produced in a certain factory. We use the word *population* to refer to all members of a group, all things with a certain description.

Sometimes we are most interested to know the typical value of the data. For example, say we are looking at household income in our town. One indicator would be the figure such that half the population earns more and half earns less. This is called the *median* income. Another, found by adding all the incomes and dividing by the number of households, is called the *mean*. These are called measures of *central tendency*. We shall look further at these measures, and others, later in this chapter.

## 4.1  Describing Data

Most data has some sort of numerical value, which we shall call a *variable*. The values of the variable are often called *readings*. In some cases, it is useful to assign a numerical value. For example, if you were polling prospective voters in an election with two candidates, Ms Smith and Mr Jones, the answers will be either "Smith" or "Jones," but it may be convenient to allocate purely artificial numerical values; we could say "Smith equals 1" and "Jones equals 0." If the mean is .45, then the poll shows that 45% of the people surveyed said they would vote for Smith.

The numerical values that describe a distribution are referred to as *parameters* of the distribution.

Suppose we have a collection of data. (Such a collection is often referred to as a *population*, and the individual items are called *observations* or *scores*.) For example, consider the scores students received for a quiz. Obviously our variable will be the score. The number of times a certain score occurs is called the *frequency* of the score, and set of all the scores, with their frequencies, is called the *distribution* of scores in the quiz.

W.D. Wallis, *Mathematics in the Real World*, DOI 10.1007/978-1-4614-8529-2_4,
© Springer Science+Business Media New York 2013

For example, suppose the highest possible score on the quiz was 10, and the 20 students scored as follows:

$$9,2,7,7,4,7,7,10,6,6,10,3,10,9,4,7,6,6,7,6.$$

For convenience, we rearrange the scores in ascending order. First, we gather the information; the diagram would look like Fig. 4.1, and is often called a *frequency table*.

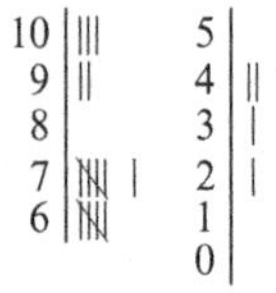

```
10 |||| 			5 |
 9 || 			4 | ||
 8 |			3 | |
 7 | ||| |		2 | |
 6 | ||||		1 |
			0 |
```

**Fig. 4.1**  Frequencies of the quiz scores

Then the list, in ascending order, is

$$2,3,4,4,6,6,6,6,6,7,7,7,7,7,7,9,9,10,10,10.$$

It is convenient to represent observations in a diagram. The two most common are the *dotplot* and the *histogram.*

In a dotplot, the possible scores are arranged on a horizontal axis and a dot is placed above the score for every observation with that value: in the example, the dotplot is

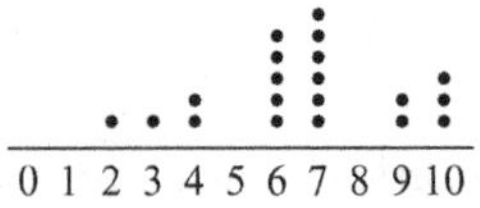

```
              :
          : : :
      . . :   : : : :
    _____
     0 1 2 3 4 5 6 7 8 9 10
```

The frequencies of the scores are

| Score     | 0 | 1 | 2 | 3 | 4 | 5 | 6 | 7 | 8 | 9 | 10 |
|-----------|---|---|---|---|---|---|---|---|---|---|----|
| Frequency | 0 | 0 | 1 | 1 | 2 | 0 | 5 | 6 | 0 | 2 | 3  |

A *histogram* is similar to a dotplot, but the dots are replaced by vertical columns. A scale on the left shows the frequencies. A histogram for the quiz scores is shown in Fig. 4.2.

A histogram is often drawn with groups of observations, called *classes*, represented in the same column. This is usually done when the number of values is large, or when there is a special meaning assigned to observations in a group. For example, if you were representing the scores in a test with a maximum score of 100, you would possibly use only 20 columns, for scores 0–4, 5–9, 10–14, and so on. In our example, there are not so many possible scores, but you might still wish to group them: for example, 8–10 is Good, 5–7 is Passing, and 2–4 is Failing. The number of

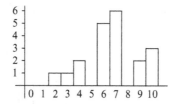

**Fig. 4.2** Histogram of the quiz scores

possible scores in a class is called the *length* of the class; usually, all groups are the same length (in our example, the length is 3).

Sometimes the columns are disjoint, as in the histogram shown in Fig. 4.3a. In other cases, the columns touch; in that case, a column whose left side is at score $a$ and whose right side is at score $b$ represents the number of scores $x$ where $a \leq x < b$. We have shown an example of this in Fig. 4.3b.

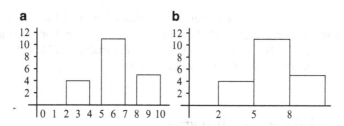

**Fig. 4.3** Histograms of the quiz scores with scores in classes

**Sample Problem 4.1** *A class took a test. The highest possible score was 20, and students scored as follows:*

$$19, 18, 8, 17, 7, 16, 10, 3, 15, 15, 13, 14, 13, 13, 15, 12, 16, 10, 9, 9, 18, 17, 7, 7, 15.$$

- (i) *How many students took the test?*
- (ii) *Draw a dotplot of the scores.*
- (iii) *Represent the scores in a histogram with classes of length 2, and in a histogram with classes of length 3.*

**Solution.**
- (i) 25 students.
- (ii) The scores, in order, are
    3, 7, 7, 7, 8, 9, 9, 10, 10, 12, 13, 13, 13, 14,
        15, 15, 15, 15, 16, 16, 17, 17, 18, 18, 19.

(iii) The histograms are:

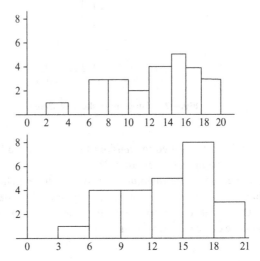

In some cases it is necessary to define the value of the variable. For example, suppose we have the set of ages of all the people in a certain town. For most purposes it is sufficient to know the age to within a year, so the variable would be the number of years in the age, rounded down to a whole number.

## 4.2 Central Tendency: Histograms

As we said, in order to answer the question, "what is the typical value of the data," we look at the measures of *central tendency*. We already mentioned two of these, the *median* and the *mean*. Another is the *mode*.

We shall refer to the number of scores in a distribution as its *size*. We do not mean just the number of scores; we take the frequency of a score into account. For example, in Worked Example 4.1 the distribution has size 25. The median is the middle value of the distribution; if a distribution has size $2n + 1$, and we list the values as $x_1, x_2, \ldots, x_{2n+1}$, where $x_1 \leq x_2 \leq \ldots \leq x_{2n+1}$, then the median is $x_{n+1}$. If the number of entries is even, a small modification is needed: the median of $x_1, x_2, \ldots, x_{2n}$, where $x_1 \leq x_2 \leq \ldots \leq x_{2n}$ is $\frac{1}{2}(x_n + x_{n+1})$.

The *mean*, or average, of a distribution is found by adding all the scores and dividing by the size. We write $\mu$ ( *mu*, the Greek letter equivalent to *m*) for the mean of a population. The *mode* is the score with the highest frequency. Sometimes there will be no mode, because the highest frequency is attained by two or more scores.

**Sample Problem 4.2** *Find the median, mean, and mode of the scores in Worked Example 4.1.*

**Solution.** The scores, in ascending order, are

$$3,7,7,7,8,9,9,10,10,12,13,13,13,14,15,15,15,15,16,16,17,17,18,18,19.$$

There are 25 scores. So the median is the 13th score, which happens to be 13. The sum of the scores is 317, so the mean is $\frac{317}{25} = 12.68$. The mode, or most common score, is 15.

**Your Turn.** Repeat the above question for the set of test scores

$$2,7,7,3,4,6,7,7,9,9,7,10,5,6,7,10,10,4,6,6.$$

A histogram is often the best tool for getting an overall picture of the distribution. In large population, the histogram often looks like a smooth curve (like the examples in Fig. 4.4, below). The highest point in the curve will be the mode; if there are two high points with a drop in between them, the distribution is called *bimodal*, but only the higher of the two high points is called a "mode."

People often expect the median to be near the center of the histogram. This is not always true. A *symmetric* distribution is one in which the part of the curve to the left of the median is roughly a mirror image of the part to the right of the median. A distribution is called *skewed* if the observations on one side of the median extend noticeably farther from the median than the observations on the other side. (We say the distribution is *skewed to the left* if the long tail is to the left, and *skewed to the right* if it is in the other direction.) When a distribution is skewed, the mean will lie nearer to the tail than the median does. For example, in most towns there are a small number of very wealthy people, and a histogram of family incomes will be skewed to the right. The mean income will be greater than the median income.

Figure 4.4 shows examples of a bimodal histogram and a histogram that is skewed to the left, from large populations.

Bimodal          Skewed to left

**Fig. 4.4** A bimodal and a skewed histogram

Some histograms will have gaps, whole groups with no entries. An *outlier* is an individual score that is separated from the rest of the distribution. For example, in the set of quiz scores in Worked Example 4.1, the score 3 is an outlier. If the histogram is symmetric, the mean and median are about the same. If it is skewed to left, the median is usually *greater* than the mean (although things like outliers can affect this). If it is skewed to right, median is usually *smaller* than the mean.

People often confuse the median and the mean. For example, a real estate agent who notes that the mean housing price for an area is $175,000 might conclude that half of the houses in the area cost more than that. The real estate agent is confusing the median of the set of data with the mean. If there are outliers in the data (two very expensive houses, with all other homes around $100,000) that can make the mean value of the homes larger, half the homes will not have prices that fall above the mean.

Note that the length of the groups can make a difference in the information we get from a histogram. For example, the outlier in Worked Example 4.1 does not show up when groups of width 3 are used. As another example, consider the following histogram; the entries are scores 1, 2, 3, . . . .

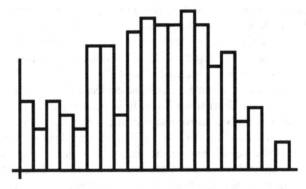

If we group the scores as 1–2, 3–4, ... we get

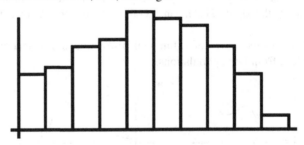

while if we group them as 1–4, 5–8, ... we get

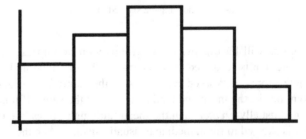

Observe the effect: we lose the fine detail, but the overall trend is easier to see.

## 4.3   Quartiles: The Boxplot

In addition to central tendency, we can measure how close a distribution is to its mean and median and how much the readings vary—what we might call the *spread* of the data.

One very simple measure is the *range* of the distribution, the amount that the largest and smallest readings vary from each other. If the minimum value is $x$ and the maximum is $y$, the word "range" is commonly used to refer either to the ordered pair $(x, y)$ or to the difference $y - x$.

More information is given by the *quartiles*. The first quartile is the value such that one-quarter of the readings are smaller than it and three-quarters are greater. To be more precise, if there are $2n$ or $2n + 1$ readings, the first quartile (or lower quartile) is the median of the set of the $n$ smallest readings. The third quartile (or upper quartile) is the median of the set of the $n$ largest readings. As an example, if there are nine readings, $\{1, 2, 3, 3, 3, 5, 5, 7, 9\}$, then the median is the fifth reading, 3. To find the first quartile, look at the set of readings preceding the median, namely, $\{1, 2, 3, 3\}$, and take its median; there are four readings, so we take the average of the second and third readings, namely 2.5. And the third quartile is $\frac{5+7}{2} = 6$.

The *five-figure summary* of a distribution is a list of five numbers—the lowest value, the first quartile, the median, the third quartile, and the highest value. The five-figure summary combines the median, the range and the quartiles. The five-figure summary of the set of readings discussed in the previous paragraph,

$$\{1, 2, 3, 3, 3, 5, 5, 7, 9\},$$

is

$$1, 2.5, 3, 6, 9.$$

A common way to display the five-figure summary of a distribution is the *boxplot*. This is a diagram drawn next to a scale with the values of the scores in the distribution. It consists of a *box*—a rectangle with one end on the lower quartile value and the other end—with tails from the box to the minimum and maximum values, and a bar at the median. For example, the boxplot for the five-figure summary $1, 2.5, 3, 6, 9$, discussed above, is

It is possible that two of the numbers in a five-number summary could be equal. If the median equals one of the quartiles, this can be shown in the boxplot by thickening the corresponding vertical line; for example, the five-figure summary $1, 3, 3, 6, 9$ could be represented as

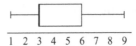

1  2  3  4  5  6  7  8  9

**Sample Problem 4.3** *Find the five-figure summary of* 1, 1, 3, 4, 4, 7, 7, 8, 8, 11, 11, 13, 14 *and represent it in a boxplot.*
**Solution.** The summary is 1, 3.5, 7, 11, 14. The boxplot is

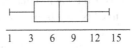

1     3     6     9    12    15

**Your Turn.** repeat for 1, 2, 3, 5, 6, 7, 9, 9, 9, 11, 13, 13, 15

Another, more technical, measure of spread is the *standard deviation*. To define this, we first define the *squared deviation* of any reading to be the square of its difference from the mean. The *variance* of a set is the average squared deviation— add all the squared deviations and divide by the order of the population. The *standard deviation* is the square root of the variance. The standard notation for a standard deviation is $\sigma$, the Greek equivalent of *s*.

**Sample Problem 4.4** *Find the variance and standard deviation of the population* $\{1, 1, 3, 3, 3, 4, 4, 5\}$.
**Solution.** The sum is 24, the order is $n = 8$, so the mean is $24/8 = 3$. We now calculate

| Reading | Deviation from mean | Squared deviation |
|---------|---------------------|-------------------|
| 1 | −2 | 4 |
| 1 | −2 | 4 |
| 3 | 0 | 0 |
| 3 | 0 | 0 |
| 3 | 0 | 0 |
| 4 | 1 | 1 |
| 4 | 1 | 1 |
| 5 | 2 | 4 |
|   |   | 14 |

The variance or average squared deviation is $14/8 = 1.75$. So the standard deviation is $\sqrt{1.75} = 1.323\ldots$ or approximately 1.23.

## 4.4   Probability Distributions

Suppose an event has a number of possible outcomes, and the theoretical probability of each outcome is known. This collection of data is called the *probability distribution* of the event. We can make a histogram of the distribution; in each case, the height of a column represents the probability that the value of the event lies in the range corresponding to the column. For example, suppose three quarters are tossed, and the number of heads is recorded. For convenience, suppose the three quarters are labeled 1, 2 and 3. There are eight possible throws: HHH, HHT, HTH, HTT, THH, THT, TTH, TTT, where the three letters represent the result (Head or Tail) of the toss of coins 1, 2 and 3 in that order. Each of these throws is equally likely, with probability 1/8, and exactly one must occur. So the probability of 0 heads is 1/8 (TTT must have occurred); one head has probability 3/8 (HTT, THT or TTH); two and three heads have probabilities 3/8 and 1/8. The histogram is

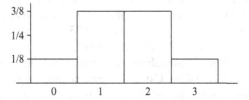

**Sample Problem 4.5**   *Supposed a biased coin has a 40% probability of landing heads on any throw. In an experiment, the coin is tossed three times and the number of heads is recorded. Find the probability distribution of the result and draw the histogram for this probability distribution.*
**Solution.** The probability of rolling the sequence $HHH$ is $0.4 \times 0.4 \times 0.4 = .064$; for $HHT$ it is $0.6 \times 0.4 \times 0.4 = .096$; and so on. The following table shows the probabilities, and the sum of probabilities for each number of heads:

| Heads | 0 | 1 | | | 2 | | | 3 |
|---|---|---|---|---|---|---|---|---|
| Rolls | $TTT$ | $HTT$ | $THT$ | $TTH$ | $HHT$ | $HTH$ | $THH$ | $HHH$ |
| Prob | .216 | .144 | .144 | .144 | .096 | .096 | .096 | .064 |
| Sum | .216 | .432 | | | .288 | | | .064 |

The histogram is

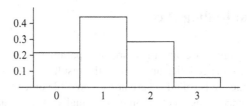

**Your Turn.** Repeat for the case where the probability of a head is 70%.

Suppose each outcome of the event has a numerical value. Then the *mean* of a probability equals the expected value. If there are $n$ possible values, $x_1, x_2, \ldots, x_n$, and the probability of value $x_i$ is $p_i$, then the mean is $\sum_{i=1}^{n} p_i x_i$. The mean is often denoted by $\mu$, the Greek letter "mu" that corresponds to *m*.

The *median* is the value such that the probability of this result or a smaller one is 50%. The *first* and *third quartiles*, $q_1$ and $q_3$, are the 25% and 75% values respectively; for example, the probability of a result smaller than or equal to the first quartile is 25%. The *interquartile range* is the difference $q_3 - q_1$. All of these quantities are analogous to the corresponding ones for samples.

In many cases, the median and quartiles are most easily found by calculating the *cumulative probability* of the event. The cumulative probability $cp(x)$ of a value $x$ is the probability that the outcome will be equal to or less than that value. That is, it is the sum of the probabilities of all outcomes with value less than or equal to $x$. The *cumulative probability distribution* of the event is defined analogously to the probability distribution, using cumulative probabilities instead of probabilities. The first quartile is the smallest value $x$ with $cp(x) \geq 0.25$; the median and third quartile are defined similarly, with 0.5 and 0.75 replacing 0.25.

The *standard deviation* of a probability distribution is defined analogously to the standard deviation of a set of data. Using the same notation as we did for the mean, the *squared deviation* corresponding to value $x_i$ is $(x_i - \mu)^2$, and the mean squared deviation, or *variance*, is $\Sigma_{i=1}^{n} p_i(x_i - \mu)^2$. The *standard deviation* equals the square root of the variance. It is commonly denoted $\sigma$ (the lower-case Greek letter "sigma," which corresponds to $s$).

In the coin-tossing example there are finitely many outcomes, so the histogram consists of a number of vertical columns. In other cases, the outcome can take any real value in a certain range. For example, the heights of individuals in the United States can take any real-number value between the heights of the shortest and tallest individuals in the country. In that sort of case, the histogram is bounded above by a curve, the *frequency curve*, and below by the vertical axis. The coin -tossing example can also be seen in this way, with the frequency curve being a sequence of horizontal and vertical line segments. In any case, the probability that the value of the outcome lies between $x$ and $y$ is precisely the area bounded by the curve, the horizontal axis, and the two vertical lines at values $x$ and $y$.

## 4.5   Examples: Rolling Dice

To show how histograms arise, and the associated parameters, we shall look at a few examples of rolling balanced dice and adding the results..

First, suppose one die is rolled. There are six possible outcomes, $\{1, 2, 3, 4, 5, 6\}$, and if the die is balanced these six outcomes are equally likely—each has probability $\frac{1}{6}$. The histogram is very simple, and is shown in Fig. 4.5. The mean is

$$\frac{1}{6} \times 1 + \frac{1}{6} \times 2 + \frac{1}{6} \times 3 + \frac{1}{6} \times 4 + \frac{1}{6} \times 5 + \frac{1}{6} \times 6 = 3.5.$$

The median is also 3.5. The first quartile is 2 (we expect that a quarter of the rolls will be 2 or smaller). The easiest way to calculate this is that there are three equally

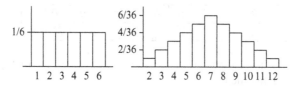

**Fig. 4.5** Histograms for rolling one or two dice

likely values less than the mean, namely 1, 2 and 3; if we know that the roll will be less than 3.5, there are three possibilities, each with probability $\frac{1}{3}$, and the median of that set is 2. Similarly the third quartile is 5.

For the variance, the calculation is

$$\frac{1}{6} \times (1 - 3.5)^2 + \frac{1}{6} \times (2 - 3.5)^2 + \frac{1}{6} \times (3 - 3.5)^2 + \frac{1}{6} \times (4 - 3.5)^2$$
$$+ \frac{1}{6} \times (5 - 3.5)^2 + \frac{1}{6} \times (6 - 3.5)^2$$
$$= \frac{1}{6} \times (6.25 + 2.25 + 0.25 + 0.25 + 2.25 + 6.25)$$
$$= \frac{17.5}{6} = 2.917 \text{ approximately},$$

and the standard deviation equals the square root of this, or approximately 1.71.

Say you roll two dice. For convenience, assume you can distinguish the two; maybe they are different colors, maybe you roll them separately. In any case, if you roll $X$ on the first die and $Y$ on the second, denote the result as $XY$. There are 36 possible rolls, 11, 12, 13, ..., 66. Each of these results is equally likely with probability $\frac{1}{36}$.

Usually, when rolling dice, you are interested in the total shown on the dice. As an example, the total 5 is obtained by the four rolls 14, 23, 32 and 41, and no others. Each has probability $\frac{1}{36}$, and they are mutually exclusive, so the probability of a roll with total 5 is $\frac{4}{36}$ or $\frac{1}{9}$. In the same way we can calculate the probability of every possible total: in the following table, Prob represents the probability of rolling a given total.

| Total | Possible Rolls | Prob | Total | Possible Rolls | Prob |
|-------|----------------|------|-------|----------------|------|
| 2 | 11 | 1/36 | 8 | 26 35 44 53 62 | 5/36 |
| 3 | 12 21 | 2/36 | 9 | 36 45 54 63 | 4/36 |
| 4 | 13 22 31 | 3/36 | 10 | 46 55 64 | 3/36 |
| 5 | 14 23 32 41 | 4/36 | 11 | 56 65 | 2/36 |
| 6 | 15 24 33 42 51 | 5/36 | 12 | 66 | 1/36 |
| 7 | 16 25 34 43 52 61 | 6/36 | | | |

The histogram is also shown in Fig. 4.5. The mean and median are easily seen to be 7. (When the histogram is symmetric about some central value, it is easy to see that both mean and median will equal that central value.) The cumulative probability distribution is

| Total $x$ | 2 | 3 | 4 | 5 | 6 | 7 | 8 | 9 | 10 | 11 | 12 |
|---|---|---|---|---|---|---|---|---|---|---|---|
| $cp(x)$ | $\frac{1}{36}$ | $\frac{3}{36}$ | $\frac{6}{36}$ | $\frac{10}{36}$ | $\frac{15}{36}$ | $\frac{21}{36}$ | $\frac{26}{36}$ | $\frac{30}{36}$ | $\frac{33}{36}$ | $\frac{35}{36}$ | 1 |

Since $6/36 = 0.17$ and $10/36 = 0.27$ to two decimal places, the first quartile is 5; similarly, the third quartile is 9.

To two decimal places, the variance comes to 5.83, and the standard deviation is 2.42.

The case of three dice is handled similarly. There are $6^3 = 216$ possible rolls. The probabilities of the different sums are

| Total | Prob | Total | Prob | Total | Prob | Total | Prob |
|---|---|---|---|---|---|---|---|
| 3 | 1/216 | 7 | 15/216 | 11 | 27/216 | 15 | 10/216 |
| 4 | 3/216 | 8 | 21/216 | 12 | 25/216 | 16 | 6/216 |
| 5 | 6/216 | 9 | 25/216 | 13 | 21/216 | 17 | 3/216 |
| 6 | 10/216 | 10 | 27/216 | 14 | 15/216 | 18 | 1/216 |

yielding a mean and median of 10.5 and quartiles 8 and 13. The variance comes to 8.75, and the standard deviation is approximately 2.96. The histogram is shown in Fig. 4.6.

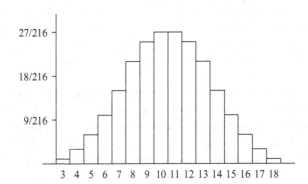

**Fig. 4.6** Histogram for rolling three dice

Finally, we consider rolling four dice. There are $6^4 = 1296$ possible rolls, and the total can range from 4 to 24.

| Total | Prob | Total | Prob | Total | Prob |
|---|---|---|---|---|---|
| 4 | 1/1296 | 11 | 104/1296 | 18 | 80/1296 |
| 5 | 4/1296 | 12 | 125/1296 | 19 | 56/1296 |
| 6 | 10/1296 | 13 | 140/1296 | 20 | 35/1296 |
| 7 | 20/1296 | 14 | 146/1296 | 21 | 20/1296 |
| 8 | 35/1296 | 15 | 140/1296 | 22 | 10/1296 |
| 9 | 56/1296 | 16 | 125/1296 | 23 | 4/1296 |
| 10 | 80/1296 | 17 | 104/1296 | 24 | 1/1296 |

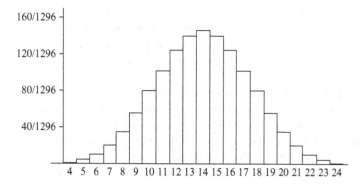

Fig. 4.7 Histogram for rolling four dice

with mean and median of 14 and quartiles 12 and 16. The variance is 11.666..., and the standard deviation is about 3.41. The histogram is shown in Fig. 4.7.

## 4.6 The Normal Distribution

In many cases, the histogram of a probability distribution looks like the bell-shaped curve shown in Fig. 4.8. In that case, the curve is called a *normal curve* and the distribution is called a *normal distribution*. The curve extends infinitely far in both directions, but the probability of an extreme value is very small.

Some simple probability distributions are close to the normal distribution. In general, when there are a number of different factors added together to give the value of the event, the distribution is close to the normal. For example, the heights of adults are approximately normal. They obviously do not follow a normal distribution exactly—the probability of a negative height is zero—but the distribution is very close to the normal. We commonly say a variable is "distributed normally" when the distribution is close to normal, even if not exactly normal. One important example is that nearly all sample means are approximately normal. After all, you calculate the mean by adding a number of different factors—the different values of the variable in the sample—before you divide by the number of members. So the properties of the normal distribution can be used in analyzing sample means.

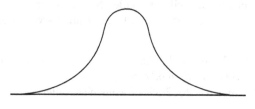

Fig. 4.8 The normal curve

The histogram of the normal distribution is a symmetric bell-shaped curve. The precise shape depends on the amount of spread, which is most commonly measured by the standard deviation. Suppose the mean is $\mu$ and the standard deviation is $\sigma$. Then the center of the curve, its highest point, is at $\mu$, the mean, which also equals the median and the mode.

In a normal distribution, approximately 68.2689% of the scores will lie in the range from $\mu - \sigma$ to $\mu + \sigma$. In other words, the probability that a reading lies between $\mu - \sigma$ and $\mu + \sigma$ is about 68%, or 0.68 if you prefer. (For our purposes, it is usually sufficient to give probabilities to the nearest percentage.) The first quartile is approximately $\mu - \frac{2}{3}\sigma$, and the third quartile is approximately $\mu + \frac{2}{3}\sigma$.

As an example, suppose the weights of 1-year-old children in your city average 25 lbs, and the standard deviation of weights is 3 lbs. (These are approximations of the figures for 1-year-olds in the United States, as given by National Health Statistics; in the real world, the standard deviation is a little larger.) The weights of children of a given age are very close to a normal distribution. So we would expect that about 68% of the 1-year-old children in the city weigh between 22 and 28 lbs. The normal histogram is symmetric about the mean, so we would expect 34% to lie between 22 and 25 lbs, and another 34% between 25 and 28. The first quartile is 23, so about one-fourth of 1-year-olds weigh less than 23 lbs, another fourth are between 23 and 25, and so on.

About 95% of the scores in a normal distribution lie within two standard deviations of the mean—between $\mu - 2\sigma$ and $\mu + 2\sigma$—so (using the symmetry) there are 2.5% of scores greater than $\mu + 2\sigma$, and the same number less than $\mu - 2\sigma$. So among the 1-year-olds, all but 5% would have weights in the range from 19 to 31 lbs.

**Sample Problem 4.6** *Assume the height of women in the United States is normally distributed with mean 64 in and standard deviation 2.4 in. Find ranges that contain the heights of the middle 68% of all United States women and the middle 95% of all women.*

**Solution.** The middle 68% of scores in a normal distribution with mean $\mu$ and standard deviation $\sigma$ lie between $\mu - \sigma$ and $\mu + \sigma$. So 68% of women's heights lie between 61.6 and 66.4 in. The formula for 95% uses $2\sigma$ instead of $\sigma$, so the range is 59.2–68.8 in.

**Your Turn.** What is the range of heights of the middle 50%?

The facts that 68% of scores lie in the range $\mu \pm \sigma$ and 95% in the range $\mu \pm 2\sigma$ is called the *68–95 rule* for normal distributions, and is illustrated in Fig. 4.9.

Since 68% or scores lie in the range $\mu \pm \sigma$, there are 32% outside that range. Half (16%) will be greater and half will be smaller. So 16% of readings lie above $\mu + \sigma$ and another 16% lie below $\mu - \sigma$. Similarly 2.5% are above $\mu + 2\sigma$ and 2.5% are below $\mu - 2\sigma$.

**Sample Problem 4.7** *Given the data in the preceding Sample Problem, how many US women are taller than 68.8 in?*

**Solution.** As 68.8 in is $\mu + 2\sigma$, 2.5% of women are taller.

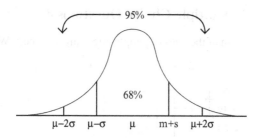

**Fig. 4.9** 68–95 rule

The percentage that lie within three standard deviations of the mean is about 99.7%, and the corresponding percentage for four standard deviations is a little greater than 99.99%.

If you look at the histograms for rolling dice that we constructed in the preceding section, you will see that the more dice you roll, the more the histogram looks like a normal curve.

# Multiple Choice Questions 4

1. Below is a histogram of the ages of people attending a concert. Which statement is true?

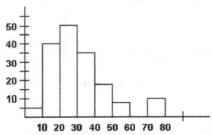

   A.   There is a gap in the histogram.
   B.   The histogram is roughly symmetric.
   C.   The histogram is skewed to the left.
   D.   The center of the distribution is at about age 50.

2. Below is a histogram of the ages of the audience at a lecture. Which statement is true?

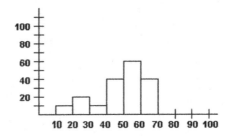

   A.   The histogram is roughly symmetric.
   B.   There is an outlier at 100.
   C.   The histogram is skewed to the left.
   D.   The histogram has a gap.

3. Here are the heights (in inches) of students in a fourth-grade class. Find the mean height.

$$39, 40, 48, 49, 40, 42, 48, 53, 47, 42, 49, 51, 52, 45, 47, 48$$

   A.   48              B.   50              C.   47              D.   46.25

4. The numbers of cats owned by families in a our neighborhood are:

$$4, 3, 7, 1, 0, 5, 1, 2, 10, 0, 0, 3$$

What is the median number of cats owned?
   A.   2              B.   2.5              C.   3              D.   4

**5.** Here are the numbers of errors made by a typist on various pages of typing:

$$14, 6, 12, 19, 2, 35, 4, 5, 3, 7, 5, 8.$$

Find the first quartile for the number of errors per page.
A.  5              B.  13              C.  15.5              D.  4.5

**6.** The shelf life of a battery produced by a major company is known to be normally distributed, with a mean life of 4 years and a standard deviation of 0.75 years. What is the upper quartile of battery shelf life?
A.  4.5 years      B.  5.2 years      C.  4.75 years      D.  5.5 years

**7.** In the preceding question, what range of years contains 68% of all battery shelf lives?
A. 2.5–5.5 years   B. 3–5 years   C. 3.25–4.75 years   D. 2.88–4.67 years

**8.** A histogram is skewed to the left. Then
A.  The mean and median are about the same.
B.  The median is greater than the mean.
C.  The mean is greater than the median.
D.  In some cases the median will be greater, in some cases the mean will be greater.

**9.** Here are the lengths (in minutes) of 12 phone calls made on an 800 line to a business on 1 day. Find the five-number summary for these data.

$$14, 6, 12, 19, 2, 35, 5, 4, 3, 7, 5, 8$$

A.  2, 4, 7, 14, 35            B.  2, 4, 6, 12, 19
C.  2, 4.5, 6.5, 13, 35        D.  5, 8, 14, 15.5, 20

**10.** The University's Distance Learning Center records how many students take an online test each week. The numbers are

$$0, 1, 3, 4, 5, 2, 2, 6, 8, 2, 0, 1, 3, 6, 3.$$

What is the five-number-summary of this data?
A.  0, 1.5, 3, 5.5, 8          B.  0, 4, 6, 1, 3
C.  0, 1.5, 3.5, 5.5, 8        D.  0, 1, 3, 5, 8

**11.** What is the standard deviation of $\{5, 15, 50, 18, 42, 36\}$?
A.  13.9          B.  3.41          C.  12.7          D.  161.33

**12.** A poll of 100 students found that 20% were in favor of raising parking fees to build a new parking garage. The standard deviation of this poll is 4%. What would be the standard deviation if the sample size was increased from 100 to 200 students (answer to one decimal place)?
A.  8%           B.  2%           C.  5.6%           D.  2.8%

**13.** Calculator batteries have useful lives that are distributed approximately normally, with an average life of 2000 h and a standard deviation of 100 h. If a random sample of 100 batteries is chosen and the mean useful life of members of the sample is calculated, what is the standard deviation of this mean?

A.   1 h                  B.   10 h                  C.   25 h                  D.   100 h

# Exercises 4

1. In each case, calculate the mean, median and (if possible) mode of the given set of data.
   (i)  1, 1, 1, 1, 1, 2, 3, 3, 4, 4
   (ii) 1,4, 2, 4, 3,4, 3
2. In each case, calculate the mean, median and (if possible) mode of the given set of data.
   (i)  1, 1, 3, 4, 1, 1, 1, 2, 3, 4
   (ii) 1, 1, 7, 7, 7, 2, 3
3. In each case, calculate the mean, median and (if possible) mode of the given set of data.
   (i)  1, 1, 2, 4, 5, 5, 3
   (ii) 12, 4, 6, 8, 3, 1, 1, 7, 5
4. In each case, calculate the mean, median and (if possible) mode of the given set of data.
   (i)  1, 1, 2, 3, 3, 3, 6, 8, 9
   (ii) 1, 1, 2, 4, 5, 5
5. In the list of five measurements shown below, one is unreadable. (The ? represents the unreadable figure.) What must this value be, if the mean of the five measurements is 6?

$$8, \ 5, \ 10, \ ?, \ 3$$

6. Here are the weights, in ounces, of squash grown in a garden one summer. Construct a frequency table using classes of width 5 ounces. Then make a histogram of the weights. Does the distribution appear to be symmetric?

| 12 |  8 | 23 | 14 | 25 | 22 | 26 | 28 | 35 | 26 |
|----|----|----|----|----|----|----|----|----|----|
| 27 | 26 | 30 | 18 | 40 | 26 | 31 | 11 | 20 | 28 |
| 30 | 29 | 36 | 22 | 38 | 32 | 26 | 12 | 17 | 18 |
| 42 | 21 | 27 |  7 | 25 | 18 | 23 | 14 | x23 | 22 |
| 28 | 11 | 20 | 32 | 24 | 18 | 12 | 32 | 19 | 38 |

7. Below are the exam scores of 30 students. Find the five-number summary of this data and make a boxplot.

| 24 | 31 | 38 | 49 | 51 | 55 | 56 | 59 | 60 | 63 |
|----|----|----|----|----|----|----|----|----|----|
| 65 | 66 | 69 | 72 | 72 | 73 | 76 | 81 | 81 | 84 |
| 85 | 86 | 86 | 88 | 88 | 88 | 91 | 91 | 92 | 97 |

8. In each case, what is the five-number summary of the given set of data? Make a boxplot of the summary.
   (i)  1, 5, 5, 3, 1, 2, 3, 3, 2, 4, 5, 2
   (ii)  1,4, 6, 6, 3, 7,4, 3
9. In each case, what is the five-number summary of the given set of data? Make a boxplot of the summary.
   (i)  1, 2, 3, 4, 5, 3, 1, 3, 4, 1, 1, 1
   (ii)  11, 13, 17, 12, 7, 12, 14
10. In each case, what is the five-number summary of the given set of data? Make a boxplot of the summary.
    (i)  21, 20, 19, 14, 13, 23, 16, 28, 11, 13, 12
    (ii)  2, 4, 6, 8, 3, 1, 1, 7, 5
11. Four fair coins are tossed and the total number of heads is recorded. Find the probabilities of zero, one, two, three and four heads and draw the histogram.
12. A biased coin with a 60% probability of landing heads is tossed three times and the number of heads is recorded. Find the probability distribution of the result and draw the histogram for this probability distribution.
13. The shelf life of a battery produced by a major company is known to be normally distributed, with a mean life of 4 years and a standard deviation of 0.75 years. What is the upper quartile of battery shelf life? What range of years contains the middle 68% of all battery shelf lives?
14. The average length of time per week that students in Jacob's class spend on homework is approximately normally distributed with a mean of 16 h and a standard deviation of 3 h. Jacob plans to spend more time on homework each week than 75% of his classmates. What is the minimum time he must spend?
15. Explain why the conclusion drawn in each case is not valid.
    (i)  A businesswoman calculates that the median cost of five business trips she took in a month is $600 and concludes that the total cost must have been $3000.
    (ii)  A company executive concludes that an accountant made an error because he prepared a report stating that 90% of the company's employees earn less than the mean salary.

# Chapter 5
# Sampling: Polls: Experiments

In many cases, we would like to know some properties of a whole population. For example, manufacturers need to know which new products are in demand; a state government would like to know which of two proposed new highways would be used more; newspapers would like to predict elections. In these cases, the cost of asking all members of a population would be prohibitive. So we check with a sample of the population, and try to predict from these results.

In other cases, examining the whole population just does not work. If a pharmaceutical company wants to know whether a medication is effective in curing a disease, they cannot test it on all cases of the disease; they are interested in treating those who will contract the disease in the future. And in some cases they wish to compare two treatments that cannot both be given to the same patient. Again, the solution is to treat a sample.

In this chapter we consider an important question: if we have results from a sample, how do we use these to predict the results for the population, and how reliable are our conclusions?

## 5.1 Predictions

Say you want to know the mean of a distribution. Your best approach is to take a sample and find its mean; you would expect the mean of your sample to be close to the mean of the distribution. The bigger the sample, the more accurate you expect the sample mean to be. In fact, the mean of a sample is itself a variable. For example, if you are sampling from a population of size $N$, and you take a sample of size $n$, there are $\binom{N}{n}$ possible samples, so the population of means is a set of $\binom{N}{n}$ numbers, and you are essentially choosing one of them at random. As we pointed out in the previous chapter, sample means approximately follow a normal distribution. If the original distribution has mean $\mu$ and standard deviation $\sigma$, we can assume that the distribution of sample means is approximately normal, with mean $\mu$ and standard deviation $\sigma/\sqrt{n}$.

W.D. Wallis, *Mathematics in the Real World*, DOI 10.1007/978-1-4614-8529-2_5,
© Springer Science+Business Media New York 2013

**Sample Problem 5.1** *Suppose you roll two dice and add the scores, 100 times. What is the expected value of the average roll? What is its standard deviation?*
**Solution.** The mean sum is 7, and the standard deviation is 2.42. So you expect the mean of the sample to be 7, and the standard deviation of this mean is 0.242.
**Your Turn.** Repeat this for three dice.

In this example, we used the population distribution to find out some properties of sample means. The reverse process is often used: when you want to know more about some variable (family incomes in an area, for example, or the heights of 9-year-olds), you take a sample. The main worry people have about sample data is: how reliable is a sample? The standard deviation is helpful here.

Suppose you measure some property; the sample mean is $m$ and you calculate the standard deviation of your sample as $s$. (We shall talk about reliable ways to estimate a population standard deviation in the next section.) Then these are your best guesses for the population mean $\mu$ and standard deviation $\sigma$. So we assume that *sample means* for this property are distributed approximately normally, with mean $m$ and standard deviation $s/\sqrt{n}$. Turning this around, our best guess for the population mean is $m$, and there is a 95% probability that the population mean lies between $m - 2(s/\sqrt{n})$ and $m + 2(s/\sqrt{n})$. This is called the *95% confidence interval* for the population mean, and $2(s/\sqrt{n})$ is the *margin of error*.

**Sample Problem 5.2** *A sample of 100 workers in a factory have a mean income of $42,000. The standard deviation of the sample is $4,000. What is the 95% confidence interval for the average salary of workers?*
**Solution.** Here $m = \$42,000, s = \$4,000$ and $n = 100$, so $2(s/\sqrt{n}) = 2 \times \$4,000/10 = \$800$ and the 95% confidence interval is $\$42,000 \pm \$800$ or $\$41,200$ to $\$42,800$.

Sometimes we know the mean of a distribution. For example, there are reliable estimates available of family incomes in districts, population weights and heights at various ages, and so on. In that case, we can tell whether a sample mean is close to the actual mean. If the population mean lies outside the 95% confidence interval calculated from the sample mean, this usually indicates bias in the sample. A smaller difference is normally attributed to variability in the population.

The measurements that are taken of a sample are called *statistics*.

## 5.2   Sampling

Often collecting all the data from a population is too expensive or too time-consuming, so we look at part of the whole collection. *Sampling* is the process for selecting a subset of members of the population. When we do not know some parameter of the population, we plan to use the relevant statistic to estimate the parameter. We want to get a *representative* sample, so that we can make reliable deductions about the population from the sample. We want to avoid a *biased* sample, one with some particular trend or tendency. For example, if you want to ask questions about smoking on campus, you would be wrong to base your conclusions on the reactions of a group who were standing outside a building, smoking.

Examples of biased sampling are common. In 1948, the New York Times announced that (Republican candidate) "Thomas E. Dewey's election as President is a foregone conclusion," and just before the election, Life Magazine ran a large picture of Dewey with the same sentence as caption. However, polls conducted in country feed stores and small movie theaters predicted Harry S. Truman as an easy winner. Harry S. Truman won the popular vote by more than two million votes and carried 303 electoral votes to Dewey's 189. The New York Times and Life predictions were based on biased polls.

When the Pacifist Brotherhood of America ran a poll to determine what Americans thought about guns and gun control, the organization was very pleased to find that 95% of those surveyed favored gun control laws. Participants for the survey were chosen from the group's newsletter mailing list. When the National Rifle Association ran a similar poll, the results were nearly 100% against. The reason for this disparity is easy to see: both organizations took polls of their memberships. Not everyone in the population in question (all adult Americans) had an equal chance of being chosen for the polls.

One way to solve this problem is *simple random sampling*, where any sample of correct size has equal chance of being the one chosen. It follows that every member of the population is equally likely to be chosen. If a polling company wants to predict an election, they might randomly choose 1,000 names from the electoral roll and ask those people how they expect to vote.

In order to make random selections, it is useful to use a *random number table*. This is an array consisting of copies of the digits 0, 1, 2, 3, 4, 5, 6, 7, 8, 9 with no pattern whatsoever. Each digit has probability $\frac{1}{10}$ of being the entry in any particular position. Figure 5.1 is an example, although tables used in practice could be much larger, spread over several pages. The entries in each row have been collected in columns of width 5, and the rows and columns of the table have been labeled for ease of reference.

Suppose your University has 8,000 students and you wish to choose a simple random sample of 50 students. First, obtain a list of all students. This will not be random—it will probably be in alphabetical order, or else divided by faculties or years. Then choose 50 numbers in the range from 1 to 8,000. This could be done by selecting ten rows—say rows 1, 3, 4, 7, 11, 12, 16, 20, 21 and 31—and five columns—say 2, 3, 4, 6 and 7. In each case, select the number formed by the first four digits in the entry. For example, row 4, column 3 results in number 1,219. Row 7, column 4 yields 0176; ignore the initial zero and the number is 176. When you have finished, delete any repeats or numbers over 8,000 and select enough new numbers to give 50 choices. Then use these numbers to select students: 1,219 means you select the 1,219th member of the student list.

|    | 1     | 2     | 3     | 4     | 5     | 6     | 7     | 8     | 9     | 10    |
|----|-------|-------|-------|-------|-------|-------|-------|-------|-------|-------|
| 1  | 65358 | 70469 | 87149 | 89509 | 72176 | 18103 | 55169 | 79954 | 72002 | 20582 |
| 2  | 72249 | 04037 | 36192 | 40221 | 14918 | 53437 | 60571 | 40995 | 55006 | 10694 |
| 3  | 48917 | 48129 | 48624 | 48248 | 91465 | 54898 | 61220 | 18721 | 67387 | 66575 |
| 4  | 88378 | 84299 | 12193 | 03785 | 49314 | 39761 | 99132 | 28775 | 45276 | 91816 |
| 5  | 77800 | 25734 | 09801 | 92087 | 02955 | 12872 | 89848 | 48579 | 06028 | 13827 |
| 6  | 24028 | 03405 | 01178 | 06316 | 81916 | 40170 | 53665 | 87202 | 88638 | 47121 |
| 7  | 86558 | 84750 | 43994 | 01760 | 96205 | 27937 | 45416 | 71964 | 52261 | 30781 |
| 8  | 78545 | 49201 | 05329 | 14182 | 10971 | 90472 | 44682 | 39304 | 19819 | 55799 |
| 9  | 30734 | 71571 | 83722 | 79712 | 25775 | 65178 | 07763 | 82928 | 31131 | 30196 |
| 10 | 64628 | 89126 | 91254 | 24090 | 25752 | 03091 | 39411 | 73146 | 06089 | 15630 |
| 11 | 42831 | 95113 | 43511 | 42082 | 15140 | 34733 | 68076 | 18292 | 69486 | 80468 |
| 12 | 80583 | 70361 | 41047 | 26792 | 78466 | 03395 | 17635 | 09697 | 82447 | 31405 |
| 13 | 00209 | 90404 | 99457 | 72570 | 42194 | 49043 | 24330 | 14939 | 09865 | 45906 |
| 14 | 05409 | 20830 | 01911 | 60767 | 55248 | 79253 | 12317 | 84120 | 77772 | 50103 |
| 15 | 95836 | 22530 | 91785 | 80210 | 34361 | 52228 | 33869 | 94332 | 83868 | 61672 |
| 16 | 14969 | 64623 | 82780 | 35686 | 30941 | 14622 | 04126 | 25498 | 95452 | 63937 |
| 17 | 58697 | 31973 | 06303 | 94202 | 62287 | 56164 | 79157 | 98375 | 24558 | 99241 |
| 18 | 38449 | 46438 | 91579 | 01907 | 72146 | 05764 | 22400 | 94490 | 49833 | 09258 |
| 19 | 62134 | 87244 | 73348 | 80114 | 78490 | 64735 | 31010 | 66975 | 28652 | 36166 |
| 20 | 72749 | 13347 | 65030 | 26128 | 49067 | 27904 | 49953 | 74674 | 94617 | 13317 |
| 21 | 81638 | 36566 | 42709 | 33717 | 59943 | 12027 | 46547 | 61303 | 46699 | 76243 |
| 22 | 46574 | 79670 | 10342 | 89543 | 75030 | 23428 | 29541 | 32501 | 89422 | 87474 |
| 23 | 11873 | 57196 | 32209 | 67663 | 07990 | 12288 | 59245 | 83638 | 23642 | 61715 |
| 24 | 41692 | 40581 | 93050 | 48734 | 34652 | 41577 | 04631 | 49184 | 39295 | 81776 |
| 25 | 61885 | 50796 | 96822 | 82002 | 07973 | 52925 | 75467 | 86013 | 98072 | 91942 |
| 26 | 13862 | 72778 | 09949 | 23096 | 01791 | 19472 | 14634 | 31690 | 36602 | 62943 |
| 27 | 08312 | 27886 | 82321 | 28666 | 72998 | 22514 | 51054 | 22940 | 31842 | 54245 |
| 28 | 39634 | 62349 | 74088 | 65564 | 16379 | 19713 | 39153 | 69459 | 17986 | 24537 |
| 29 | 14595 | 35050 | 40469 | 27478 | 44526 | 67331 | 93365 | 54526 | 22356 | 93208 |
| 30 | 11071 | 44430 | 94664 | 91294 | 35163 | 05494 | 32882 | 23904 | 41340 | 61185 |
| 31 | 82509 | 11842 | 86963 | 50307 | 07510 | 32545 | 90717 | 46856 | 86079 | 13769 |
| 32 | 07426 | 67341 | 80314 | 58910 | 93948 | 85738 | 69444 | 09370 | 58194 | 28207 |
| 33 | 57696 | 25592 | 91221 | 95386 | 15857 | 84645 | 89659 | 80535 | 93233 | 82798 |
| 34 | 24965 | 54524 | 82780 | 35631 | 70441 | 14622 | 64623 | 04126 | 25498 | 63937 |
| 35 | 09262 | 25041 | 57862 | 19203 | 86103 | 02800 | 23198 | 70639 | 43757 | 52064 |

**Fig. 5.1**  A table of random integers

The process described would be very tedious, but computers can be used to do the steps. There are also random number generating programs available.

Of course, the numbers in the table will not really be random; they were generated by a computer, and if you know the program you could presumably predict the number in any position. But for practical purposes, the numbers behave as random.

Another method is *stratified sampling*, where the sampling organization tries to select proportionally from the population. In the voting example, suppose a city voted 35% Republican, 25% Democrat in the last election; 40% did not vote. In order to poll 1,000 voters, a polling company might wish to survey 350 chosen from those who voted Republican last time, 250 Democrats and 400 who did not vote.

## 5.3 Interpreting Samples

The main reason for sampling is to get an idea of the properties of the whole population. For example, if we want to know the mean of a population, our best guess is the mean of a representative sample, and similarly for the median. Naturally, a bigger sample will lead to a more reliable estimate.

Measures of spread in a sample can be used to estimate the spread of a population, but be careful! When you calculate the standard deviation of a sample, you use deviations from the mean of the *sample*, not the *population*. The sample mean is the best estimate of the population mean, but it is not perfect. The best estimate of the population mean would be made by using the squared deviations from the population mean, not the sample mean; but in most cases this is not available. When the population mean is not known, it can be shown that the best estimate of the population variance is found by summing the squared deviations from the sample mean, and then dividing by $n - 1$, where $n$ is the sample size. Usually we write $m$ for the sample mean, and $s$ for the estimate of the standard deviation found by dividing the sum of the squared deviations from $m$ by $n - 1$.

That is, if the readings in a sample are $x_1, x_2, \ldots, x_n$, then the *sample* standard deviation is

$$\sqrt{\frac{\sum (x_i - \bar{x})^2}{n}},$$

but the best estimate of the *population* standard deviation is

$$\sqrt{\frac{\sum (x_i - \bar{x})^2}{n - 1}}.$$

To get a 95% confidence interval for the mean of a normal (or approximately normal) population (such as human attributes like height and weight, or family incomes for large groups like cities), one takes a sample—let's say its mean is $\bar{x}$—and use

$$\bar{x} \pm \frac{2s}{\sqrt{n}}$$

where, if the population standard deviation is known, $\sigma$ say, we use $s = \sigma$; but if the population standard deviation is not known, we use

$$s = \sqrt{\frac{\sum (x_i - \bar{x})^2}{n - 1}}.$$

**Sample Problem 5.3** *You observe the output of a machine for 4 h. The machine produces the following number of screws per hour:*

$$hour\ 1 : 20,050 screws$$

$$hour\ 2 : 19,990 screws$$

$$hour\ 3 : 20,020 screws$$

$$hour\ 4 : 19,980 screws.$$

*Assuming these are typical hours of production, and assuming the hourly production follows approximately a normal distribution, find a 95% confidence interval for hourly screw production.*
**Solution.** The mean is

$$(20,050 + 19,990 + 20,020 + 19,980)/4 = 20,010.$$

The sum of squared deviations is

$$40^2 + 20^2 + 10^2 + 30^2 = 3,000.$$

So the standard deviation estimate is

$$\sqrt{\frac{3,000}{3}} = \sqrt{1,000} = 31.625$$

and the 95% range is

$$20,010 \pm 63.25.$$

In any kind of sampling, it is important to avoid bias, if we want to use the results to make predictions about the whole population.

Perhaps the most common type of bias in sampling is the *convenience sample*. For example, an interviewer might wait outside a supermarket door and ask customers a question as they arrive. The customers at a supermarket might not be representative of the whole population. Another example is the telephone survey, which is biased toward those who have their telephones handy.

Another example of bias is self-selection. We have all seen surveys in newspapers and on television where the public are invited to respond. This type of survey will over-represent those who feel strongly about the topic. Another form of self-selection is when opinions are sought from a sample but only those who choose to respond are counted. This is called a *voluntary response sample.*

Sometimes the survey itself is biased. The question itself might suggest an answer.

A common form of sampling is an *observational study.* These are often conducted to test drugs, production methods, and so on. For example, to test whether too much chocolate increases the chance of diabetes, 100 rats were fed one pound of chocolate a day for 6 months. Fifteen percent developed diabetes, while the percentage among rats is only 3%. This does not prove that too much chocolate causes diabetes; other factors might be the increased calories in the diet, lack of exercise, laboratory conditions, and so on. This interference from other factors is called *confounding.*

**Sample Problem 5.4** *A sociologist wants to know the opinions of employed adult women about government funding for day care. She obtains a list of the 520 members of a local business and professional women's club and mails a questionnaire to 100 of these women selected at random. Only 48 questionnaires are returned. What are the population and the sample in this survey? Is this survey biased?*

**Solution.** The population is all employed adult women in the area represented by the club, and the sample is the set of 100 women randomly chosen. The study is almost certainly biased; business and professional women are not representative of all employed women, and the fact that only those who chose to respond are included will very probably affect the result.

To avoid confounding, a comparative study, also called an *experimental* study, is usually conducted. In this case other factors are controlled. The experimental subjects are divided into two groups, the *experimental* group and the *control* group. Both groups receive exactly the same treatment except for the variable being tested.

For example, in the rat experiment, one could use 200 rats instead of 100. The original set of rats is the *experimental* group, while the new 100 are the *control* group. Other factors (laboratory conditions, calorie intake) are the same for the two groups; only the one variable—the amount of chocolate—varies between the two groups.

This technique works well when testing inanimate objects, or rats, but what if the experimental subjects are people? For example, suppose a new drug is being tested. The very fact that you know you are receiving a drug, or that you know there is a new drug available and you are not receiving it, could influence your reaction.

For this reason, *placebos* are used. A placebo is a fake drug, looking exactly the same as the actual drug but not containing the active ingredient. For example, if the tablets used in an experiment contain 5 g of a new hormone substitute, the placebo might instead contain 5 g of sugar. (This is a common example, and even if sugar and salt are not involved, placebo drugs are often called "sugar pills" or "salt pills.") Subjects in medical experiments have a tendency to respond favorably to any treatment, whether the real one or the placebo; this is called the "placebo effect."

In experiments involving placebos, it is standard practice that neither the participants nor those who test for effects are told which treatment was received. In particular, if the subject is being interviewed about the effect of treatment, an interviewer who knew the difference might ask different questions depending on whether the subject received the test treatment or the placebo. The way to avoid this is for the interviewer not to know whether or not the particular subject is in the control group. Tests conducted in this way are called *double blind*.

## 5.4   Polls

Suppose the University wants to estimate satisfaction with a new course. They might choose a sample of 200 students and ask them whether they enjoyed the course. Say 46 of the students say they enjoyed it. These 46 students are 23% of the sample, so we call 23% the *sample proportion*. The percentage of students who actually enjoyed the course is called the *population proportion*. Assuming the sample of 200 students was unbiased, the University's best guess for the population proportion is 23%.

In this example, a number of people were asked to give an opinion. This is called an *opinion poll*. These are very often used to evaluate products, test public opinion of political candidates, and so on. A random sample is chosen and the sample proportion is found. This is the best guess as to the population proportion, and of course it is important to know how confident one can be of the result.

Assume we are discussing a proportion (such as "proportion voting for a candidate" or "proportion voting yes") and estimating it by sampling. We shall write $p$ for the population proportion, and $\hat{p}$ ("p-hat") for the corresponding sample proportion. It will be convenient always to write these as percentages.

If you count 1 for each "yes" and 0 for each "no," $p$ is the population mean and $\hat{p}$ is the sample mean. So $\hat{p}$ has approximately a normal distribution, and $\hat{p}$ is the best estimator of $p$.

If the sample size is $n$, then $\hat{p}$ has mean $p$ and standard deviation $\sigma_p$, where it can be shown that approximately

$$\sigma_p = \sqrt{\frac{p(100 - p)}{n}}.$$

(This approximate standard deviation is quite accurate for large $n$, for example if $n$ is at least 1,000.)

The best estimate for this standard deviation is

$$s_p = \sqrt{\frac{\hat{p}(100 - \hat{p})}{n}}.$$

So we assume that there is a 95% chance that $p$ lies in the range from $\hat{p} - 2s_p$ to $\hat{p} + 2s_p$.

**Sample Problem 5.5**  *In a survey of 1600 voters, 960 say they intend to vote Democrat at the next election. What is the 95% confidence interval for the percentage of voters who will vote Democrat?*
**Solution.** In this case $n = 1600$ and

$$\hat{p} = \frac{960}{1600} = 60\%.$$

Therefore

$$s_p = \sqrt{\frac{\hat{p}(100 - \hat{p})}{n}}$$

$$= \sqrt{\frac{60(100 - 60)}{1600}}$$

$$= \sqrt{\frac{60 \times 40}{1600}}$$

$$= \sqrt{\frac{3}{2}} = 1.225 \text{ approximately.}$$

The 95% confidence interval is

$$\hat{p} - 2s_p \text{ to } \hat{p} + 2s_p$$

that is

$$60 - 2.45 \text{ to } 60 + 2.45$$

or

$$57.55\% \text{ to } 62.45\%.$$

(This is often written as $60 \pm 2.45\%$.)
**Your Turn.** Approximately 27% of all the students who go to Fort Lauderdale get sunburned during spring break. A poll of 500 vacationers will be taken and each asked if they were sunburned.
   (i)  What is the expected number who say yes?
   (ii) What is the standard deviation of the sampling distribution of the percentage who will say yes?
   (iii) What is a 95% confidence interval for this percentage?

If there had been only 100 people in the survey, $s_p$ would have been 4.9, and the 95% confidence interval would have been

$$50.2\% \text{ to } 69.8\%.$$

—not nearly so accurate.

There is another way of thinking about the 95% confidence level, which may give you a better idea of what it means. Suppose you went through the process of finding the 95% confidence interval for every possible sample. Then in 95% of cases the true population parameter value (mean, proportion, or whatever) would lie in the interval.

Notice that the size of the confidence interval depends on the parameter (the percentage voting yes, for example) and on the sample size, but *not* on the population size. In the example, the sample size 1600 was equally reliable if there were 18,000 voters or 180,000,000. This may go against your intuition, but it is a fact.

## 5.5  Latin Squares

The combination of several experimental variables can make experiments prohibitively large and expensive. However, combinatorics, the mathematical study of arrangements, can sometimes provide clever arrangements of treatments that help hold down the size and cost of experiments. We will illustrate this by a comparison of motor oils.

Some makers of motor oil claim that using their product improves gasoline mileage in cars. Suppose you wish to test this claim by comparing the effect of four different oils on gas mileage. However, because the car model and the driver's habits greatly influence the mileage obtained, the effect of an oil may vary from car to car and from driver to driver. To obtain results of general interest, you must compare the oils in several different cars and with several different drivers. If you choose 4 car models and 4 drivers, there are 16 car–driver combinations. The type of car and driving habits are so influential that you must consider each of these 16 combinations as a separate experimental unit. If each of the 4 oils is used in each unit, $4 \times 16$, or 64, test drives are needed.

However, there is a way to test the effects of cars, drivers, and oils that requires only 16 test drives. Call the oils $A, B, C,$ and $D$ and assign one oil to each of the 16 car–driver combinations in one of the two arrangements shown in the figure below. The different rows represent the different cars and the different columns represent the different drivers. Study these arrangements: each oil appears 4 times, exactly once in each row (for each car) and also exactly once in each column (for each driver). This setup can, in fact, show how each oil performs with each car and with each driver.

$$
\begin{array}{cccc}
A & B & C & D \\
B & A & D & C \\
C & D & B & A \\
D & C & A & B
\end{array}
\qquad\qquad
\begin{array}{cccc}
A & B & C & D \\
C & D & B & A \\
B & A & D & C \\
D & C & A & B
\end{array}
$$

An arrangement like that in the motor oil example is called a *Latin square experiment*, and the figure is called a $4 \times 4$ *Latin square*; it arranges 4 kinds of objects (the oils) in a $4 \times 4$ matrix. The "Latin" property is that each label appears exactly once in each row and in each column. There are several different $4 \times 4$ Latin squares. In a Latin square experiment, one first chooses a Latin square of the needed size at random from all the possible squares. Then the four drivers are assigned at random to the columns, the four cars at random to the rows, and the four motor oils at random to the labels $A, B, C, D$. After the test drives are completed, the results are handed over to a statistician, who has ways to test whether the results obtained might have been obtained by chance, or show a real difference in the effect of the oils.

In general an $n \times n$ Latin square is a square array with $n$ rows, $n$ columns, whose entries are chosen from $n$ symbols. Each symbol occurs exactly once in each row and exactly once in each column.

Here is a $5 \times 5$ example:

$$
\begin{array}{ccccc}
1 & 2 & 3 & 4 & 5 \\
2 & 1 & 4 & 5 & 3 \\
3 & 4 & 5 & 1 & 2 \\
5 & 3 & 1 & 2 & 4 \\
4 & 5 & 2 & 3 & 1
\end{array}
$$

You have probably seen Latin squares in the form of sudoku puzzles. A standard *sudoku array* is a $9 \times 9$ Latin square partitioned into nine $3 \times 3$ subsquares—rows 1,2,3 and columns 1,2,3, rows 1,2,3 and columns 4,5,6, rows 1,2,3 and columns 7,8,9, rows 4,5,6 and columns 1,2,3, and so on, up to rows 7,8,9 and columns 7,8,9— with the property that every subsquare contains every symbol exactly once. In the puzzle, certain entries are omitted, and you have to find them; usually the array is designed so that there is exactly one answer.

A sudoku array need not be $9 \times 9$; puzzle books often include (easier) examples of size $6 \times 6$ with $2 \times 3$ subsquares or (harder) $16 \times 16$ examples with $4 \times 4$ subsquares. The right-hand $4 \times 4$ square shown above is a (very easy) sudoku square, a $4 \times 4$ square with $2 \times 2$ subsquares.

# Multiple Choice Questions 5

1. A chocolate factory wishes to check on the actual weight of its 100 g bittersweet chocolate bars. A sample of 100 bars is taken and the mean weight is 99.2 g, with a standard deviation of 0.5 g. What is the 95% confidence interval for the mean weight of bars produced by the factory?

    A.  99.15–99.25 g                B.  99.1–99.3 g
    C.  98.7–99.7 g                 D.  98.2–100.2 g

2. Given the sample $\{25, 15, 50, 18, 42, 36\}$, what is the best estimate of the population standard deviation?

    A.  13.9         B.  3.41        C.  12.7        D.  161.33

3. A random sample of 20 pages of a textbook was made and 60% of the pages had illustrations on them. What is the 95% confidence interval for the actual proportion of pages on which an illustration would be found.

    A.  29–91%                 B.  38.1–81.9%
    C.  44.5–75.7%               D.  49.1–70.9%

4. Your University wishes to survey the students to determine their feelings about the quality of services in the student cafeteria. Which of the following sampling designs is best for avoiding bias?
    A.  Place an ad in the student newspaper asking readers to mail in their opinions.
    B.  Obtain a list of student names from the registrar and select 250 names to contact.
    C.  Run an announcement on the campus radio station asking listeners to phone in their opinions.
    D.  Survey every tenth student who enters the Cafeteria.

5. A flashlight manufacturer sets aside a production line for the assembly of 1000 flashlights to fill a special order. Fifty of these flashlights are selected at random from the production line to be tested, and 12 are found to be defective. The population is
    A.  the 12 defective flashlights.
    B.  the 50 flashlights tested.
    C.  the 2000 flashlights produced for the special order.
    D.  all flashlights produced by the manufacturer.

6. Random selection of subjects for surveys is used to avoid

    A.  bias                      B.  the placebo effect
    C.  double blindness           D.  variability

7. A poll of 1000 voters finds that 36% prefer candidate Brown. This 36% is a

    A.  parameter    B.  population    C.  statistic        D.  sample

8. In order to determine the proportion of voters in a small town who favor a candidate for mayor, the campaign staff takes out an ad in the paper asking voters to call in their preference for mayor. This type of sample is a

    A.  convenience sample           B.  simple random sample
    C.  stratified sample              D.  voluntary response sample

9. To determine the proportion of students at a university who would enroll in a new course, 200 members are randomly chosen from a list of prospective students and surveyed. It is found that 70 would select the new course. What is the sample proportion?

    A.  3.5%         B.  7%          C.  35%         D.  70%

10. The Jackson County Fire Department conducted a survey to determine how the 60,000 residents of the county feel about the fire department's ability to meet their needs. A random sample of 720 citizens was selected, and 637 of them said that they feel confident about the fire department's service. The population is

    A.  the 60,000 residents of Jackson County.
    B.  the 720 citizens surveyed.
    C.  the 637 confident citizens.
    D.  the employees of the Jackson County Fire Department.

# Exercises 5

1. In your town, records show that the probability of rain on a random day in April is 20%. How many rainy days would you expect next April?

2. A machine makes wood screws with an average length of 4 in and a standard deviation of .01 in. What is the 95% confidence interval for the length of these screws?

3. The mean weight of a collection of pumpkins in a shipment to a fruit market is 3.3 pounds, with a standard deviation of 0.4 pounds. The distribution of weights is approximately normal. What is the probability that one pumpkin chosen at random will weigh more than 3.7 pounds?

4. The distribution of the duration of human pregnancies (that is, the number of days between conception and birth) has been found to be approximately normal with a mean of 266 days and a standard deviation of 16 days. Determine the proportion of all pregnancies that last:
    (i)  between 250 and 282 days;
    (ii)  between 234 and 298 days;
    (iii)  less than 234 days.

5. Consider the set of readings

$$12, 14, 16, 17, 16, 18, 14, 16, 18, 20, 19, 12.$$

    (i)  Find the mean, variance, and standard deviation of this set.
    (ii)  If these readings are a sample from a larger population, what are your estimates for the mean and standard deviation of the population?

6. Home canners sometimes preserve vegetables in used mayonnaise jars to avoid buying special canning jars. Organic Gardening magazine wondered what percent of mayonnaise jars would break when used for canning. It obtained 100 mayonnaise jars from a local supplier and canned tomatoes in them. Only three of the jars broke.

(i)  What is the population of this study?

(ii)  What is the sample of this study?

7.  In a study of the smoking habits of urban American adults, the New York Times asks 500 people their age, place of residence, and how many cigarettes they smoke daily.

(i)  What is the population of this study?

(ii)  What is the sample of this study?

8.  Your company takes a sample of 200 people to test whether they would buy a new product. The company is unhappy because the 95% confidence interval is large, and wants an interval half the size. How many people would they have to sample?

9.  A survey shows that the 95% confidence interval of the percentage of PBS listeners in your town who like the local news announcer is 76–88%. What is the mean number who like the announcer? What is the margin of error of the survey?

10.  In a certain town, family incomes are distributed normally with a mean of $48,000 and a standard deviation of $8,000. A sample of 16 families is surveyed. What is the 95% confidence interval for the mean income of this sample?

11.  A marketing department surveys 1000 shoppers and finds that 650 would visit a new store more often if were open Sunday evenings. What is the sample proportion in this survey?

12.  The army reports that the mean size of soldiers' heads (circumference) is 22.8 in and the standard deviation is 1.1 in.

(i)  What percentage of soldiers have head circumference greater than 23.9 in?

(ii)  State a 95% confidence interval for soldiers' head circumferences.

13.  There are 100 members in your local dramatic society, 40 men and 60 women. You wish to sample their opinion as to whether your next production should be a comedy or a drama. You choose six of the women at random and choose four of the men at random and ask their opinions.

(i)  Every member has the same probability of being interviewed; what is that probability?

(ii)  Why is this not a simple random sample?

14.  To estimate the height of adult women in a town, a sample of 150 women was chosen and the average height was found to be 66 in. A later complete survey found that the actual average height was 63 in, and the standard deviation of heights was 2 in. Is this difference likely to be an example of sample variability or sampling bias?

15.  A local marketing research company wishes to find out the percentage of shoppers in a town who would buy a larger size of laundry detergent if it were available. Previous surveys in other towns suggest that the percentage of shoppers who would buy the larger size will be approximately 65%. The company wishes the standard deviation of the sampling distribution for the percentage who say yes in their survey to be about 2.5%. How many shoppers should they survey?

**16.** A survey of 900 voters show that 657 favor a certain candidate. What is the sample proportion in favor of the candidate? What is the 95% confidence interval for the true proportion in favor of the candidate, to within 0.1%?

**17.** Construct a $3 \times 3$ Latin square and a $6 \times 6$ Latin square.

**18.** Officials at a university want to determine the percentage of students who favor the allocation of a percentage of tuition funds toward the construction of a new campus parking garage. To find this out, a survey is conducted. One thousand people driving through the administration building parking lot are surveyed, and it is found that 75% of these people favor the garage.

Write a memo to the officials, pointing out that this is a convenience sample. Explain why this conclusion might not be valid, with specific reference to this survey.

**19.** Give two examples of measurements where you would expect the histogram to be rather like a normal curve and one example of a measurement whose histogram you would expect to be nothing like a normal curve.

# Chapter 6
# Graphs: Traversing Roads

This chapter introduces an area of mathematics called *graph theory*. We start with the application for which graph theory was first developed, in the eighteenth century.

## 6.1 Representing Roads: The Königsberg Bridges

Suppose your vacation is to be a road trip, visiting a number of towns by car. You will want to know the roads joining various towns in order to plan your itinerary.

Of course, road atlases are available, but if your only desire is to list the towns that will be visited, in order, you do not need to know all the information in the atlas, such as the physical properties of the region—hills and so on—or whether different roads cross or there are overpasses. The important information is whether or not there is a road joining a given pair of towns. For this purpose it is sufficient to make a diagram as shown in Fig. 6.1a, that indicates roads joining $B$ to $C$, $A$ to each of $B$ and $C$, and $C$ to $D$, with no direct roads joining $A$ to $D$ or $B$ to $D$.

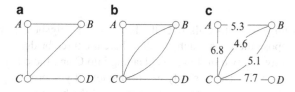

**Fig. 6.1** Representing a road system

There will often be a choice of routes between two towns. This information may be significant: you might prefer the freeway, which is faster, or you might prefer the more scenic coast road. In this case the diagram does not contain enough information. For example, if there had been two different ways to travel from $B$ to

W.D. Wallis, *Mathematics in the Real World*, DOI 10.1007/978-1-4614-8529-2_6,
© Springer Science+Business Media New York 2013

*C*, this information could be represented as in the diagram of Fig. 6.1b. Then you could go further, and label the different roads with their estimated driving times, as shown in Fig. 6.1c.

**Sample Problem 6.1** *Suppose the road system connecting towns A, B and C consists of two roads from A to B, one road from B to C, and a bypass road directly from A to C. Represent this road system in a diagram.*

**Solution.**

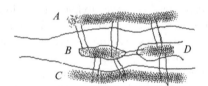

$$A \qquad B \qquad C$$

**Your Turn.** Draw a diagram to represent the road system connecting towns *A*, *B*, *C*, *D*, with one road from *A* to *B*, one from *A* to *C*, two from *A* to *D*, and one from *B* to *C*.

The sort of diagram shown in Fig. 6.1a is called a *linear graph*, or usually just a *graph*. The dots representing towns will be called *vertices*, while the lines representing the road links are called *edges*.

The study of graphs goes back to 1735, when the great Swiss mathematician Leonhard Euler gave a talk to the St. Petersburg Academy. The talk grew out of a famous old problem. The town of Königsberg in Prussia was built on the river Pregel. The river divided the town into four parts, including an island called The Kneiphof, and in the eighteenth century the town had seven bridges; the layout is shown in Fig. 6.2. The question under discussion was whether it is possible from any point on Konigsberg to take a walk in such a way as to cross each bridge exactly once.

**Fig. 6.2** The Königsberg bridges

In fact, it is impossible to walk over the bridges of Königsberg as required. To see this, let's suppose there was such a walk. There are three bridges leading to the area *C*: you can traverse two of these, one leading into C and the other leading out, at one time in your walk. There is only one bridge left: if you cross it going into *C*, then you cannot leave *C* again, unless you use one of the bridges twice, so *C* must be the finish of the walk; if you cross it in the other direction, *C* must have been the start of the walk. So either *C* is the place where you started or it is the place where you finished—let's call it an *end* of the walk.

We can make a similar analysis of parts *A, B* and *D*, since each has an odd number of bridges—five for *B* and three for the others. So each of *A, B, C, D* is either the start or the finish of the walk. But any walk starts at one place and finishes at one place; it

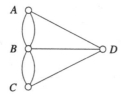

**Fig. 6.3** Graph representing the Königsberg bridges

can start and finish at the same place, or there can be a start and a separate finish. But in either case there cannot be more than two ends; $A, B, C$ and $D$ cannot all be ends.

Euler started by finding a graph that models the Königsberg bridge problem (considered as a road network, with the islands and riverbanks as separate "towns"). Vertices $A, B, C$ and $D$ represent the parts $A, B, C$ and $D$ of Königsberg, and each bridge is represented by an edge. The graph is shown in Fig. 6.3. In terms of this model, the original problem becomes: can a walk be found that contains every edge of the graph precisely once?

> **Sample Problem 6.2** *The islands A, B, C, D are joined by seven bridges—two joining A to B, two from C to D, and one each joining the pairs AC, AD and BD. Represent the system graphically. Could one walk through this system, crossing each bridge exactly once?*
>
> **Solution.** Islands $B$ and $C$ each have an odd number of bridges, so one must be the start of the walk and the other the finish. With a little experimentation you will find a solution—one example is *BACDABDC*, and there are others.
>
>
>
> **Your Turn.** The islands $X, Y, Z, T$ are joined by seven bridges—three joining $X$ to $Y$, and one each joining the pairs $XZ$, $XT$ and $YZ$. Represent the system graphically. Could one walk through this system, crossing each bridge exactly once?

Euler set for himself the more general problem: given any configuration of river, islands and bridges, find a general rule for deciding whether there is a walk that covers each bridge precisely once. Before we discuss his work, it will be useful to look at some graph-theoretic ideas and definitions.

## 6.2 Some Graph Terminology

We now make a few formal definitions, and set up some notation. A graph $G$ consists of a finite set $V$ or $V(G)$ of objects called *vertices*, together with a collection $E(G)$ of pairs of vertices that are called *edges*. (This is consistent with the use of "vertex" and "edge" earlier.) The two vertices belonging to an edge are called its *endpoints*; we say they are *adjacent*, and say the edge joins them. An edge with endpoints

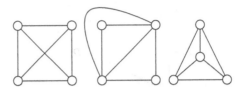

**Fig. 6.4** Three representations of the same graph

$X$ and $Y$ will be written as $\{X,Y\}$, or simply $XY$ when no confusion can arise. Adjacent vertices are called *neighbors*. We write $x \sim y$ to mean that vertices $x$ and $y$ are adjacent.

In the diagram representing a graph, an edge $XY$ is shown as a line from $X$ to $Y$. This is usually drawn as a straight line segment, but not always; the only important thing is that it joins the correct points. Moreover, the position of the points is not fixed. For these two reasons, one graph can give rise to several drawings that look quite dissimilar. For example, the three diagrams in Fig. 6.4 all represent the same graph. Although the two diagonal lines cross in the first picture, their point of intersection does not represent a vertex of the graph.

It is not clear whether or not the definition of a graph allows for two edges with the same endpoints. Some authors interpret "collection of pairs" to mean that repeated pairs are permitted, while others do not. However, there are some situations where we wish to represent several different links between two vertices. The road system modeled in Fig. 6.1b is an example. The lines representing the two roads from $B$ to $C$ in that figure will be called *multiple* edges. We say there is a multiple edge of multiplicity 2 joining $B$ to $C$. If multiple edges are present, the diagram is called a *multigraph*; if we want to emphasize the fact that multiple edges are not allowed, we use the term *simple graph*. We might also want an edge that goes from a vertex back to the same vertex, representing for example a crescent in a road system, this would be called a *loop*; the usual interpretation is that loops are not allowed, unless the term "looped graph" or "looped multigraph" is used.

Another question is, can you have a graph with no vertices at all? Some authors say "yes," and admit the concept of a graph with no vertices or edges at all. But we shall always require that a graph have at least one vertex. So the smallest possible graph will have one vertex and no edges. We shall call it the *null* graph or *empty* graph.

We shall define a *walk* $(A,B,C,\ldots,Y,Z)$ to be a sequence of vertices $(A,B,C,\ldots,Y,Z)$ such that $AB,BC,\ldots,YZ$ are all edges. For example, the diagram of Fig. 6.1a represents a graph that contains a walk $(A,B,C,D)$, because each of $AB$, $B,C$ and $C,D$ is an edge of the graph. Similarly, there is a walk $(A,C,D)$. However, there is no walk $(A,B,D)$, because there is no edge $BD$. In terms of the model as a road system, a walk is a sequence of roads that you could traverse in order. A *circuit* (or *closed walk*) is a walk that finishes at its starting point; Fig. 6.1a contains a circuit $(A,B,C,A)$. A walk is called *simple* if no edge is repeated; in graph (a), $(A,B,C,A,B)$ is a walk, but edge $AB$ is repeated, so it is not simple. The *length* of a walk is the number of edges that it contains.

The *degree* (or *valency*) of a vertex is the number of edges with that vertex as an endpoint; we'll write $d(x)$ for the degree of a vertex $x$. A vertex whose degree is odd is called an *odd vertex*; others are *even*. (For completeness, we say that each loop adds 2 to the degree of the vertex.)

**Sample Problem 6.3** *Find the degrees of all the vertices in these graphs:*

**Solution.**

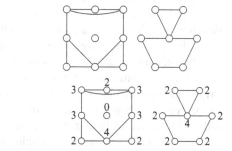

**Your Turn.** Find the degrees of all the vertices in these graphs:

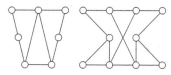

Given a graph with $e$ edges, suppose you wrote down a list of the edges, once each. Then replace each edge by its pair of vertices. For example, if $y_1$ is an edge with endpoints $x_4$ and $x_7$, you might make erase the entry $y_1$ and instead write down $x_4$ and $x_7$. Then a vertex $x$ will appear precisely $d(x)$ times in the list, so the total number of entries equals the sum of the degrees of the vertices. On the other hand, each edge appeared once in the original list, so each edge gives rise to two entries in the revised list. So there are $2e$ entries in the new list—an even number. Putting it another way, the number of edges in a graph is precisely half the sum of all the degrees.

If you add together an odd number of odd integers, together with any number of even numbers, you get an odd number. This can't have happened in the revised list. So the number of odd integers in the revised list is even. That is, in any graph, the number of odd vertices is even.

So far, nothing we have said implies that a graph must be all one piece. Figure 6.4 showed three graphs, but you could also consider them as one graph, made up of three separate parts. In that case, we would say the graph is *disconnected*. For example, Fig. 6.4 might represent the major highways on three islands that are not connected by bridges.

If the graph is all one piece, it is called a connected graph. The connected parts of a disconnected graph are called its *components*.

In a number of cases, we consider graphs in which every pair of vertices is joined by an edge. In that case, we say the graph is *complete*. Complete graphs are particularly important when a number, such as a distance or a cost, is associated with each edge. A complete graph with $n$ vertices is denoted $K_n$; for example, Fig. 6.4 shows three different ways to draw the graph $K_4$. Clearly $K_1$, the null graph, is a (rather trivial) complete graph.

An alternative way of representing a graph is its *incidence array*. This is an array in which the rows and columns both represent the vertices of the graph, in the same order. A 1 in position $(i, j)$ means that there is an edge joining the vertices, while a 0 indicates no edge. In the case of a multigraph, some people replace the 1 by the number of edges, while others simply use the array to show which vertices are connected. In cases where only the structure is important, the *incidence matrix*, with the names of the vertices not shown, contains all the information needed.

If we reorder the list of vertices, this will produce a reordering of the rows and columns of the incidence array and of the incidence matrix. Two incidence matrices are called *equivalent* when one can be obtained from the other by performing a permutation on the rows and the same permutation on the columns.

Graphs with the same structure are called *isomorphic*. Graphs are isomorphic if and only if their incidence matrices are equivalent.

## 6.3   Back to Königsberg

Instead of a configuration of islands and bridges, let us discuss the graph corresponding to it. We'll say the graph is *traversable* if we can find a walk that goes over every edge precisely once, and refer to such a walk as an *Euler walk* in the graph. We also say a graph is *Eulerian* if it has such a walk. A closed Euler walk is also called an *Euler circuit*.

In our discussion of the Königsberg bridges, we observed that an odd vertex could be at the start of the walk or at the finish. So, if a graph is traversable, it can have at most two odd vertices. You can't have an odd number of odd vertices, so there can be either two odd vertices (one will be the start, the other the finish), or none (the walk begins and ends at the same vertex). We'll refer to the walk as closed if the start equals the finish, and open otherwise.

So, if a graph has an Euler walk, all the degrees of its vertices or all but two of them must be even. Another obvious condition is that the graph must be connected, or else no single walk could cover all the edges. In fact, Euler showed that these necessary conditions are also sufficient. Formally, his result can be stated as:

*If a connected graph has no odd vertices, then it has an Euler walk starting from any given point and finishing at that point. If there are two odd vertices, then there is an Euler walk whose ends are the odd vertices.*

To see how his reasoning works, we consider any simple walk in a graph that starts and finishes at the same vertex. If one erases every edge in that walk, one deletes two edges touching any vertex that was crossed once in the walk, four edges touching any vertex that was crossed twice, and so on. (For this purpose, count "start" and "finish" combined as one crossing.) In every case an *even* number of edges is deleted.

First, consider a graph with no odd vertex. Select any vertex $x$, and select any edge incident with $x$. Go along this edge to its other endpoint, say $y$. Then choose any other edge incident with $y$. In general, on arriving at a vertex, select any edge incident with it that has not yet been used, and go along the edge to its other endpoint. At the moment when this walk has led into the vertex $z$, where $z$ is not $x$, an odd number of edges touching $z$ has been used up (the last edge to be followed, and an even number previously). Since $z$ is even, there is at least one edge incident with $z$ that is still available. Therefore, if the walk is continued until a further edge is impossible, the last vertex must be $x$—that is, the walk is closed. It will necessarily be a simple walk and it must contain every edge incident with vertex $x$.

Now assume that a connected graph with every vertex even is given, and a simple closed walk has been found in it by the method just described. Delete all the edges in the walk, forming a new graph. From the preceding paragraph it follows that every vertex of the new graph is even. It may be that we have erased every edge in the original graph; in that case we have already found an Euler walk. If there are edges still left, there must be at least one vertex, say $c$, that was in the original walk and that is still on an edge in the new graph—if there were no such vertex, then there could be no connection between the edges of the walk and the edges left in the new graph, and the original graph must have been disconnected. Select such a vertex $c$, and find a closed simple walk starting from $c$. Then unite the two walks as follows: at one place where the original walk contained $c$, insert the new walk. For example, if the two walks are

$$x, y, \ldots, z, c, u, \ldots, x$$

and

$$c, s, \ldots, t, c,$$

then the resulting walk will be

$$x, y, \ldots, z, c, s, \ldots, t, c, u, \ldots, x.$$

(There may be more than one possible answer if $c$ occurred more than once in the first walk. Any of the possibilities may be chosen.) The new walk is a closed simple walk in the original graph. Repeat the process of deletion, this time deleting the newly formed walk. Continue in this way. Each walk contains more edges than the preceding one, so the process cannot go on indefinitely. It must stop: this will only happen when one of the walks contains all edges of the original graph, and that walk is an Euler walk.

Finally, consider the case where there are two odd vertices $p$ and $q$ and every other vertex is even. Form a new graph by adding an edge $pq$ to the original (it might duplicate an existing edge, or it might not). This new graph has every vertex even. Find a closed Euler walk in it, choosing $p$ as the first vertex and the new edge $pq$ as the first edge. Then delete this first edge; the result is an Euler walk from $q$ to $p$.

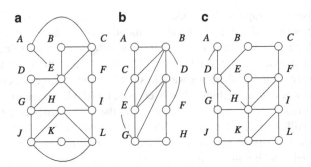

**Fig. 6.5** Find Euler walks

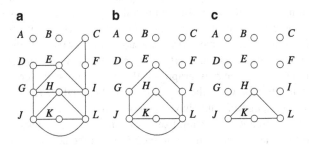

**Fig. 6.6** Constructing an Euler walk

**Sample Problem 6.4** *Find Euler walks in the road networks represented by Fig. 6.5a and* b.

**Solution.** In the first example, starting from *A*, we find the walk *ACBEA*. When these edges are deleted (see Fig. 6.6a) there are no edges remaining through *A*. We choose *C*, a vertex from the first walk that still has edges adjacent to it, and trace the walk *CFIHGDEC*, after which there are no edges available at *C* (see Fig. 6.6b). *E* is available, yielding walk *EGJLIE*. As is clear from Fig. 6.6c, the remaining edges form a walk *HJKLH*.

We start with *ACBEA*. We replace *C* by *CFIHGDEC*, yielding

$$ACFIHGDECBEA,$$

then replace the first *E* by *EGJLIE*, with result

$$ACFIHGDEGJLIECBEA$$

(we could equally well replace the second *E*). Finally *H* is replaced by *HJKLH*, and the Euler walk is

$$ACFIHJKLHGDEGJLIECBEA.$$

In the second example, there are two odd vertices, namely $B$ and $F$, so we add another edge $BF$ and make it the first edge used. The first walk found was

$$BFHGCABFDBCEB,$$

and the second is $DEGD$, exhausting all the vertices and producing the walk

$$BFHGCABFDEGDBCEB.$$

The first edge (the new one we added) is now deleted, giving Euler walk

$$FHGCABFDEGDBCEB$$

in the original graph.

**Your Turn.** Find an Euler walk in the road network represented by Fig. 6.5c.


## 6.4 Applications: Eulerization

Suppose a highway inspector needs to inspect the roads in your neighborhood. She needs to travel along every road, but will only need to go once along each. The obvious technique is to model the road system with a graph—in this case, vertices will represent intersections, and every road is shown as an edge—and find an Euler circuit in the graph. The same method can be used if you have to plan a route for a snow plow.

But suppose the graph contains no Euler walk. Then the highway inspector must repeat some edges of the graph in order to return to the starting point. We shall define an *Eulerization* of a graph $G$ to be a graph, with a closed Euler walk, that is formed from $G$ by duplicating some edges. A *good* Eulerization is one that contains the minimum number of new edges, and this minimum number is the *Eulerization number* eu(G) of G. For example, if two adjacent vertices have odd degree, you could add a further edge joining them. This would mean that the inspector must travel the road between them twice.

What if the two odd vertices were not adjacent? One new edge will not suffice—it would be the same as requiring that a new road be built! In most applications this is not feasible.

**Sample Problem 6.5** *Consider the multigraph G of Fig. 6.7. What is eu(G)? Find an Eulerization of the road network represented by G that uses the minimum number of edges.*

**Solution.** Look at the multigraph as shown on the left in Fig. 6.8. The black vertices have odd degree, so they need additional edges. As there are four black vertices, at least two new edges are needed; but obviously no two edges will

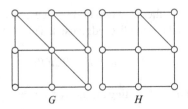

**Fig. 6.7**   Find Eulerizations

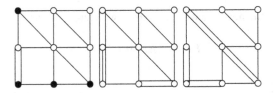

**Fig. 6.8**   Finding an Eulerization

suffice. However, there are solutions with three added edges—two examples are shown—so $eu(G) = 3$.

**Your Turn.** Consider the graph $H$ of Fig. 6.7. What is $eu(H)$? Find an Eulerization of the road network represented by $H$ that uses the minimum number of edges.

Usually edges have a cost associated with them, and the cost of an Eulerization would equal the sum of the costs of the repeated edges. The problem of finding the cheapest Eulerization is called the Chinese Postman Problem. (The first mathematician to suggest it was Chinese, publishing in a Chinese journal, and he posed it in terms of a postman's delivery route.)

For our purposes, we shall assume all edges are equal in cost. So we'll assume that the best Eulerizations are the ones with the fewest added edges.

# Multiple Choice Questions 6

**1.** Which of the following graphs are connected?

| | |
|---|---|
| A.  *G* only | B.  *H* only |
| C.  Both *G* and *H* | D.  Neither *G* nor *H* |

**2.** Which of the following graphs are connected?

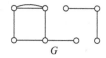   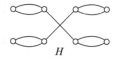

| | |
|---|---|
| A.  *G* only | B.  *H* only |
| C.  Both *G* and *H* | D.  Neither *G* nor *H* |

**3.** Which of the graphs in Question 1 have Euler walks?

| | |
|---|---|
| A.  *G* only | B.  *H* only |
| C.  Both *G* and *H* | D.  Neither *G* nor *H* |

**4.** Which of the graphs in Question 2 have Euler walks?

| | |
|---|---|
| A.  *G* only | B.  *H* only |
| C.  Both *G* and *H* | D.  Neither *G* nor *H* |

**5.** What is the degree of vertex *x* in graph *G* of Question 1?

| | | | |
|---|---|---|---|
| A.  2 | B.  3 | C.  4 | D.  5 |

**6.** What is the degree of vertex *y* in graph *H* of Question 1?

| | | | |
|---|---|---|---|
| A.  2 | B.  3 | C.  4 | D.  5 |

**7.** Which of the following graphs have Euler walks?

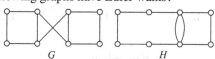

| | |
|---|---|
| A.  *G* only | B.  *H* only |
| C.  Both *G* and *H* | D.  Neither *G* nor *H* |

**8.** Which of the following graphs have Euler walks?

   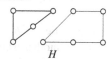

| | |
|---|---|
| A.  *G* only | B.  *H* only |
| C.  Both *G* and *H* | D.  Neither *G* nor *H* |

**9.** Which statement is true of the following graph?

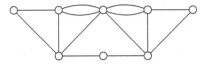

A. It has an Euler walk and an Euler circuit.
B. It has an Euler walk but no Euler circuit.
C. It has an Euler circuit but no Euler walk.
D. It has neither an Euler walk nor an Euler circuit.

**10.** Which statement is true of the following graph?

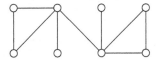

A. It has an Euler walk and an Euler circuit.
B. It has an Euler walk but no Euler circuit.
C. It has an Euler circuit but no Euler walk.
D. It has neither an Euler walk nor an Euler circuit.

**11.** What is the minimum number of edges needed to Eulerize the following graph?

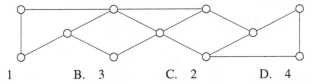

A. 1      B. 3      C. 2      D. 4

**12.** What is the minimum number of edges needed to Eulerize the following graph?

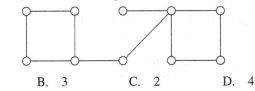

A. 1      B. 3      C. 2      D. 4

## Exercises 6

**1.** Draw graphs to represent the following road systems.
   (i) There are four cities, $A,B,C,D$. There is exactly one road joining $A$ to each of $B,C,D$. There are two roads from $B$ to $C$, and one from $B$ to $D$.
   (ii) There are six cities, $A,B,C,D,E,F$. Cities $A,B,C$ are on one island, while $D,E,F$ are on another island. There is one road joining each pair of cities on the same island, and one road (a bridge) from $B$ to $D$.
   (iii) There are five cities, $A,B,C,D,E$. Cities $A,B,C$ and $D$ are each joined to each other by single roads; $E$ is not joined to any other city.

2. Draw graphs to represent the following road systems.
   (i) There are five cities, $A, B, C, D, E$. There are two roads each joining $A$ to $B$ and $C$ to $D$, one road from $A$ to $E$ and one road from $C$ to $E$.
   (ii) There are four cities, $A, B, C, D$ joined by a ring road that goes $A$ to $B$ to $C$ to $D$ to $A$. There is a fifth city, $E$, joined to each of the other four by single direct roads.
   (iii) There are six cities $A, B, C, D, E, F$. There are roads from each of $A$ and $F$ to all four other cities, and also roads from $B$ to $C$ and from $D$ to $E$.

3. Suppose $G$ is any graph. The *square* of $G$ is a graph $G^2$ with the same vertices as $G$, and an edge between two vertices $A$ and $B$ whenever $G$ contains a walk of one or two edges from $A$ to $B$. (The square never has loops or multiple edges.) Draw the squares of the following graphs:

   (i) 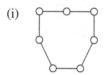   (ii)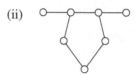

4. See the definition in the preceding question. Draw the squares of the following graphs:

   (i)    (ii)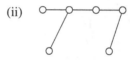

   In Exercises 5–8, find the degrees of the vertices in the given graphs.

5. (i) 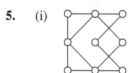   (ii)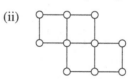

6. (i)    (ii)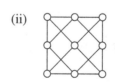

7. (i) 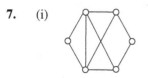   (ii)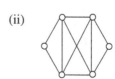

**8.**   (i)               (ii)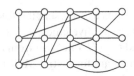

In Exercises 9–17, how many odd vertices do the given graphs contain? In each case decide whether the graph contains an Euler walk. If it does, find such a walk. Is the walk an Euler circuit?

**9.**   (i) 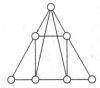              (ii)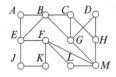

**10.**   (i) 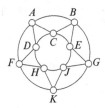             (ii)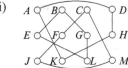

**11.**   (i)             (ii)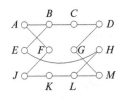

**12.**   (i)             (ii)

**13.**   (i)             (ii)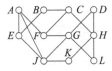

**14.** (i)

(ii)

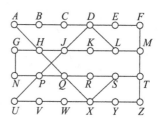

**15.** (i)

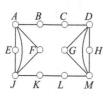

(ii)

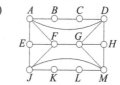

**16.** (i)

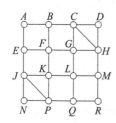

(ii)

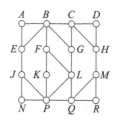

**17.** (i)

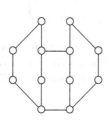

(ii)

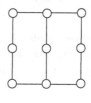

In Exercises 18–24, state the Eulerization number of the given graphs and find best-possible Eulerizations of each.

**18.** (i)

(ii)

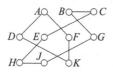

**19.** (i)

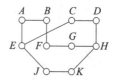

(ii)

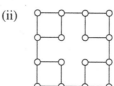

**20.** (i) 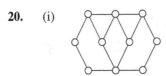    (ii)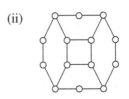

**21.** (i) 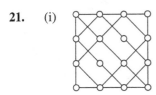    (ii)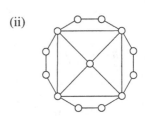

**22.** (i) 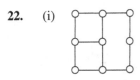    (ii)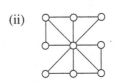

**23.** (i) 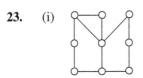    (ii)

**24.** (i) 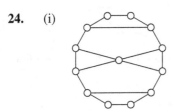    (ii)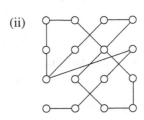

**25.** You say that a graph with odd vertices cannot have an Euler circuit. Your friend says you are wrong, and shows you the walk represented by the sequence of numbered edges on the graph below. Explain why this particular walk is not an Euler circuit.

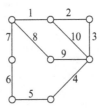

**26.** A circuit with exactly three edges is called a *triangle* in a graph. Find all triangles in the graph of Exercise 15, part (i).

**27.** Repeat the preceding exercise for the graph of Exercise 16, part (i).

**28.** Draw two graphs with six vertices, all vertices of degree 2. Your graphs should have no loops or multiple edges. One graph should be connected; the other graph should be disconnected.

**29.** Suppose you have a connected graph, and there is an edge with the property that, if you delete it from the graph, the new graph you obtain will be disconnected. Such an edge is called a *bridge*.

   (i) Find all bridges in the following graphs:

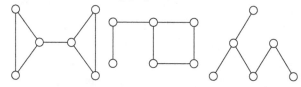

   (ii) Show that a graph with an Euler circuit cannot contain a bridge.

   (iii) Show that a graph with a bridge must contain a vertex of odd degree.

**30.** An airline executive wants his representative to check the quality of service on all the routes. Explain to him how an efficient Eulerization might be helpful.

**31.** Why would a city street department want its snow plow operator's path to follow an Euler circuit if possible?

# Chapter 7
# Graphs: Visiting Vertices

Again, let us consider graphs used to model networks of roads. In the preceding chapter, we looked at efficient ways to examine the edges of a graph—the sort of work that a highway inspector or snow plow driver might do. Now let us consider the vertices. For example, a traveling salesman might want an efficient route to visit all the cities in an area. He would not care whether all the roads were traversed. So he is concerned with vertices, not edges.

## 7.1 Some Classes of Graphs

In this chapter we will only be concerned with simple graphs; there will be no loops or multiple edges.

When discussing Euler walks and circuits, there was no restriction on the vertices; one could pass through a vertex as many times as necessary. We are now going to limit ourselves to visiting vertices once, so we need some more definitions. We shall define a *path* $(A,B,C,\ldots,Y,Z)$ to be a walk $(A,B,C,\ldots,Y,Z)$ in which there are no repeated vertex. If $(A,B,C,\ldots,Y,Z)$ is a path of length at least 2 and there is also an edge $ZA$ then the configuration consisting of the path together with the edge $ZA$ is called a *cycle*. So a cycle is a circuit with no repeated edges.

Paths and cycles can both be considered as graphs in their own right. We define the path $P_n$ to be a graph with $n$ vertices and $n-1$ edges that constitute a path of length $n-1$. For example, if the vertices are $X_1, X_2, \ldots, X_n$, the edges might be $X_1 X_2$, $X_2 X_3, \ldots, X_{n-1} X_n$. Similarly, the cycle $C_n$ is a graph whose $n$ vertices and $n$ edges form a cycle of length $n$: typically, the vertices may be $X_1, X_2, \ldots, X_n$, and the edges are $X_1 X_2, X_2 X_3, \ldots, X_{n-1} X_n, X_n X_1$. In a cycle, every vertex has degree 2.

Figure 7.1 shows copies of $P_4$ and $C_5$.

We have already seen complete graphs; remember that $K_n$ is a graph with $n$ vertices in which every pair of vertices is joined by an edge. Another interesting

W.D. Wallis, *Mathematics in the Real World*, DOI 10.1007/978-1-4614-8529-2_7,
© Springer Science+Business Media New York 2013

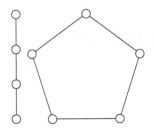

**Fig. 7.1**  A path and a cycle

graph is the *complete bipartite graph* $K_{m,n}$. This graph has two sets of vertices, one of size $m$ and one of size $n$, and two vertices are joined by an edge if and only if they are in different sets.

Another type of graph that will be important is a *tree*. This is a connected graph that contains no cycles. The tree diagrams we saw in Chap. 3 are in fact trees; and we shall see more of them in a later chapter.

By a *subgraph* we mean a collection of the vertices and edges in a graph that by themselves constitute a graph. If an edge $XY$ is included in the set of edges of a subgraph, then both vertex $X$ and vertex $Y$ will necessarily be in its set of vertices; however, if $X$ and $Y$ are vertices in a subgraph, then it is possible that $XY$ is not an edge in the subgraph, even if $XY$ was an edge in the original graph.

Figure 7.2 shows a graph $G$ and three of its subgraphs: a cycle $G_1$ (a copy of $C_3$), a path $G_2$ (a copy of $P_4$), and another subgraph $G_3$, which is a tree. Note that the cycle $G_1$ could be denoted $(A,B,E)$, $(B,E,A)$, $(E,A,B)$, $(A,E,B)$, $(E,B,A)$ or $(B,A,E)$; a cycle has no intrinsic start or finish point, and a copy of $C_n$ can be written (as a list of vertices) in $2n$ ways. On the other hand, to specify a path, you must start at one of its endpoints; $G_2$ can be written either as $(B,C,D,E)$ or as $(E,D,C,B)$.

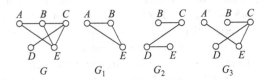

**Fig. 7.2**  Examples of subgraphs

**Sample Problem 7.1**  *In the graph G of Fig. 7.2, list all the paths that start with edge AE.*

**Solution.** $(AE)$ is itself a path, of length 1. After $A,E$ one could go on to $B$, giving the paths $(A,E,B)$, $(A,E,B,C)$, $(A,E,B,C,D)$, or to $C$, giving $(A,E,C)$, $(A,E,C,B)$, $(A,E,C,D)$.

**Your Turn.** List all the paths in $G$ that start with edge $AB$.

## 7.2 Hamilton Cycles

Suppose a traveling salesman or tourist wants to visit the towns shown on a map. They would wish to travel from one town to another, trying not to pass through any town twice on the trip, and usually they would wish to return to the starting point at the end of the trip. In terms of the underlying graph that represents the map, they would like to follow a cycle.

A cycle that passes through every vertex in a graph is called a *Hamilton cycle* and a graph with such a cycle is called *Hamiltonian*. The idea of such a spanning cycle was simultaneously developed by Hamilton in 1859 in a special case, and more generally by Kirkman in 1856.

A *Hamilton path* is a path that passes through every vertex. If you have a Hamilton cycle, you can construct a Hamilton path by deleting any one edge.

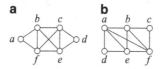

**Fig. 7.3** Two sample graphs for discussing Hamilton cycles

**Sample Problem 7.2** *Consider the graph in Fig. 7.3a. Which of the following are Hamilton cycles?*

(i) $(a,b,e,d,c,f,a)$      (ii) $(a,b,e,c,d,e,f,a)$
(iii) $(a,b,c,d,e,f,a)$      (iv) $(a,b,c,e,f,a)$

**Solution.** (i) and (iii) are Hamiltonian. (ii) is not; it contains a repeat of $e$. (iv) is not; vertex $d$ is omitted.

**Your Turn.** Repeat this problem for the graph in Fig. 7.3b and cycles

(i) $(a,b,c,f,e,d,a)$      (ii) $(a,b,f,c,b,e,d,a)$
(iii) $(a,b,c,d,e,f,a)$      (iv) $(a,d,e,b,c,f,a)$

At first, the problem of deciding whether a graph is Hamiltonian sounds similar to the problem of Euler circuits. However, the two problems are strikingly different in one regard. We found a very easy test for the Eulerian property, but no nice necessary and sufficient conditions are known for the existence of Hamilton cycles.

It is easy to see that the complete graphs with three or more vertices are Hamiltonian, and any ordering of the vertices gives a Hamilton cycle. We can discuss Hamiltonicity in a number of other particular cases, but there are no known theorems that characterize Hamiltonian graphs.

Suppose you want to find all Hamilton cycles in a graph with $v$ vertices. Your first instinct might be to list all possible arrangements of $v$ vertices and then delete those with two consecutive vertices that are not adjacent in the graph. This process can then be made more efficient by observing that the $v$ lists $a_1a_2\ldots a_{v-1}a_v$, $a_2a_3\ldots$ $a_va_1, \ldots$, and $a_va_1\ldots a_{v-2}a_{v-1}$ all represent the same cycle (written with a different starting point), and also $a_va_{v-1}\ldots a_2a_1$ is the same Hamilton cycle as $a_1a_2\ldots a_{v-1}a_v$ (traversed in the opposite direction).

Another shortcut is available. If there is a vertex of degree 2, then any Hamilton cycle must contain the two edges touching it. If $x$ has degree 2, and suppose its two neighbors are $y$ and $z$, then $xy$ and $xz$ are in every Hamilton cycle. So it suffices to delete $x$, add an edge $yz$, find all Hamilton cycles in the new graph, and delete any that do not contain the edge $yz$. The Hamilton cycles in the original graph are then formed by inserting $x$ between $y$ and $z$.

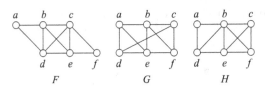

**Fig. 7.4**  Find all Hamilton cycles

**Sample Problem 7.3**  *Find all Hamilton cycles in the graph F of Fig.* 7.4.
**Solution.** The graph $F$ has two vertices of degree 2, namely $a$ and $f$. When we replace these vertices by edges $bd$ and $ce$, we obtain the complete graph with vertices $a,b,c,d$ (multiple edges can be ignored). This complete graph has three Hamilton cycles: there are six arrangements starting with $b$, namely $bcde$, $bced$, $bdce$, $bedc$, $bdec$, $becd$, and the latter three are just the former three written in reverse. $bcde$ yields a cycle that does not contain edges $bd$ and $ce$. The other two give the cycles $bcfeda$ and $badcfe$, so these are the two Hamilton cycles in $F$.
**Your Turn.** Repeat this problem for graphs $G$ and $H$ of Fig. 7.4.

## 7.3  The Traveling Salesman Problem

Suppose a traveling salesman wishes to visit several cities. If the cities are represented as vertices and the possible routes between them as edges then, as we have already said, the salesman's preferred itinerary is a Hamilton cycle in the graph.

In most cases there is a cost associated with every edge. Depending on the salesman's priorities, the cost might be a dollar cost (such as airfare or gasoline), or the number of miles to be driven, or the number of hours the trip will take. The most desirable itinerary will be the one for which the sum of costs is a minimum. The problem of finding this cheapest Hamilton cycle is called the Traveling Salesman Problem.

We shall continue to speak in terms of a salesman, but these problems have many other applications. They arise in airline and delivery routing and in telephone routing. More recently they have been important in manufacturing integrated circuits and computer chips, and for internet routing. A Hamilton cycle is often the solution to a problem where you need to examine a number of sites, while an Euler circuit is appropriate when you need to examine the routes between the sites.

In order to solve a Traveling Salesman Problem on $n$ vertices, your first thought might be to list all Hamilton cycles in the graph and then work out the cost; the cheapest answer would be the solution. But very often we can assume the graph is complete. In that case listing all the cycles can be a very long task, because of the following theorem.

**Theorem 1.** *The complete graph $K_n$ contains $(n-1)!/2$ Hamilton cycles.*

**Proof.** $K_n$ has $n$ vertices, so there are $n!$ different ways to list the vertices in order. As we pointed out previously, each cycle of length $n$ gives rise to $2n$ different ways to list its vertices. So there are $n!/(2n) = (n-1)!/2$ Hamilton cycles.          □

This number grows very quickly. For $n = 3, 4, 5, 6, 7$ the value of $(n-1)!/2$ is 1, 3, 12, 60, 360; in $K_{10}$, there are 181,440, and in $K_{24}$, there are about $10^{23}$ Hamilton cycles. Twenty-four vertices is not an unreasonably large network, but performing so many summations and comparing them would be impossible in practice. To give you some idea of the times involved, if you had a computer capable of evaluating and sorting through a million ten-vertex cycles per second, a complete search solution of the Traveling Salesman Problem for $K_{10}$ would take about .18 s. No problem so far. However, assuming the computer took about twice as much time to process a 24-vertex cycle as it took for a ten-vertex cycle, so that it could sort through half a million 24-vertex cycles per second, the complete search for $K_{24}$ would take about a billion years.

## 7.4  Algorithms for the Traveling Salesman Problem

An algorithm is a step-by-step procedure to calculate a result. We shall first introduce an algorithm, called the *brute force algorithm*, to use in those cases where a complete search is feasible. The method involves constructing a tree, called a *search tree*. Search trees are much the same as the tree diagrams we introduced in Chap. 3.

Suppose we want to find all Hamilton circuits in a graph $G$. We construct a tree, which we shall call $T$.

First, select any vertex of $G$ as a starting point; call it $V$ say. Construct the first vertex (called the *root*) of the search tree; label it with $V$. For every edge of $G$ that contains $V$, draw an edge of the tree starting at $V$ (called a *branch*) and label it with the corresponding vertex of $G$.

To illustrate this, look at the example $G$ shown on the left of Fig. 7.5. Let us start at vertex $A$. In the graph there are three vertices joined to $A$, namely $B$, $C$ and $D$. So we start three branches, and label them as shown in the right-hand side of figure.

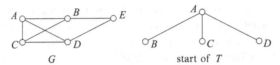

**Fig. 7.5**  The brute-force algorithm, step 1

From each new vertex, draw an edge of the tree corresponding to each edge of $G$ that could be an edge of the Hamilton cycle: it must not duplicate an edge of $G$ or complete a cycle in $G$ that misses some vertices. In the example, a branch will go from $B$ to vertices labeled $C$ and $E$, but not to $A$ (duplication of edge $AB$) or $D$ (there is no edge in $BD$ in $G$). Similarly there are two branches from each of $C$ and $D$. The result is shown in Fig. 7.6.

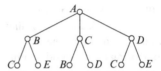

**Fig. 7.6**  The brute-force algorithm, step 2

Continue in this way from each new vertex of the tree. For example, the path $A, B, E$ can extend only to $D$, because $B$ would be a repeated edge and $E$ is not connected to $A$ or $C$. The path $A, C, D$ can extend only to $E$, because if you try to extend it to $C$ you would repeat the edge $CD$, and extending to $A$ would form a cycle of length 3, not big enough for a Hamilton cycle.

As we go on, some of the branches cannot be continued. After $A, B, C, D, E$ there is no possible continuation, because $EA$ is not an edge of $G$. Others, like $A, C, B, E, D, A$ can be completed: the branch contains every vertex of $G$ and finishes back at $A$—not a *small* circuit, because all the vertices are present. The final tree appears in Fig. 7.7.

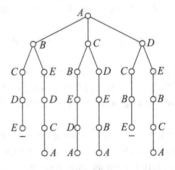

**Fig. 7.7**  The brute-force algorithm, final result

When the process is finished, the completed branches are the Hamilton cycles. Each cycle will occur twice, once in each direction. So the example has two Hamilton cycles: *ABEDCA* and *ACBEDA*.

As we observed, a complete search is not practical in large cases, especially where the graph is nearly complete. For this reason, fast methods of reaching a "good solution" have been developed. Although they are not guaranteed to give the optimal answer, these approximation algorithms often give a route that is significantly cheaper than the average. We shall look at two of these.

The *nearest-neighbor* method works as follows. Starting at some vertex *x*, one first chooses the edge incident with *x* whose cost is least. Say that edge is *xy*. Then an edge incident with *y* is chosen in accordance with the following rule: if *y* is the vertex most recently reached, then eliminate from consideration all edges incident with *y* that lead to vertices that have already been chosen (including *x*), and then select an edge of minimum cost from among those remaining. This rule is followed until every vertex has been chosen. The cycle is completed by going from the last vertex chosen back to the starting position *x*. This algorithm produces a cycle in the complete graph, but not necessarily the cheapest one, and different solutions may come from different choices of initial vertex *x*.

The *sorted-edges* method does not depend on the choice of an initial vertex. One first produces a list of all the edges in ascending order of cost. At each stage, the cheapest edge is chosen with the restriction that no vertex can have degree 3 among the chosen edges, and the collection of edges contains no cycle of length less than *v*, the number of vertices in the graph. This method always produces a cycle, and it can be traversed in either direction.

**Sample Problem 7.4**  *Suppose the costs of travel between St. Louis, Nashville, Evansville, and Memphis are as shown in dollars on the left in Fig. 7.8. You wish to visit all four cities, returning to your starting point. What routes do the two algorithms recommend?*

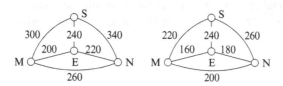

**Fig. 7.8**  A Traveling Salesman Problem example

**Solution.** The nearest-neighbor algorithm, applied starting from Evansville, starts by selecting the edge EM, because it has the least cost of the three edges incident with E. The next edge must have M as an endpoint, and ME is not allowed (one cannot return to E, it has already been used), so the cheaper of the remaining edges is chosen, namely MN. The cheapest edge originating at N is NE, with cost $220, but inclusion of this edge would lead back to E, a vertex that has already been visited, so NE is not allowed, and similarly NM is not available. It follows that NS must be chosen. So the algorithm finds route EMNSE, with cost $1040.

A different result is achieved if one starts at Nashville. Then the first edge selected is NE, with cost $220. The next choice is EM, then MS, then SN, and the resulting cycle NEMSN costs $1060.

If you start at St. Louis, the first stop will be Evansville ($240 is the cheapest flight from St. Louis), then Memphis, then Nashville, the same cycle as the Evansville case (with a different starting point), costing $1040. From Memphis, the cheapest leg is to Evansville, then Nashville, and finally St. Louis, for $1060—the same cycle as from Nashville, in the opposite direction.

To apply the sorted-edges algorithm, first sort the edges in order of increasing cost: EM($200), EN($220), ES($240), MN($260), MS ($300), NS($340). Edge EM is included, and so is EN. The next choice would be ES, but this is not allowed because its inclusion would give degree 3 to E. MN would complete a cycle of length 3 (too short), so the only other choices are MS and NS, forming route EMSNE (or ENSME) at a cost of $1060.

However, in this example, the best route is ENMSE, with cost $1020, and it does not arise from the nearest-neighbor algorithm, no matter which starting vertex is used, or from the sorted-edges algorithm.

**Your Turn.** A new cut-rate airline offers the fares shown on the right side of Fig. 7.8. What do the algorithms say now?

The preceding example illustrates the fact that the two algorithms we have presented do not necessarily give the best-possible answer. However, both algorithms give quite good solutions in most cases.

You should also realize that there is no reason to expect the cycle obtained from the nearest-neighbor algorithm to be better or worse than the result of sorted edges; sometimes one will be better, sometimes the other, and sometimes they will be the same.

When the number of vertices is relatively small, the brute-force algorithm is always best. However, it is not practical in larger examples, which often arise in the real world.

# Multiple Choice Questions 7

1. Which of the following describes a Hamiltonian cycle for the graph below?

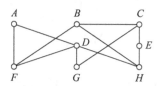

   A. *ADGCBEHFA*                        B. *AFBHECGDA*
   C. *ADHECGDFA*                       D. *ADFBCEHDA*

2. To solve the traveling salesman problem for six cities by brute force, how many different Hamilton circuits must be considered?

   A.  60           B.  120          C.  360          D.  720

3. Which of the following statements about Traveling Salesman algorithms is true?

   A.   The sorted-edges algorithm always gives as good a result as the nearest-neighbor algorithm.

   B.   The nearest-neighbor algorithm always gives as good a result as the sorted-edges algorithm.

   C.   The two algorithms always give the same result.

   D.   Sometimes one is better, sometimes the other, sometimes they are equal.

4. There are three roads from town A to other towns; they are of lengths 13, 15 and 18 miles. Using the nearest-neighbor algorithm starting from town A, which road would be chosen first?

   A.   The 13 mile road     B.   The 15 mile road     C.   The 18 mile road

   D.   There is not enough information to answer this

5. Which of the following graphs are trees?

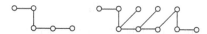

   A.  Only the left-hand graph            B.  Only the right-hand graph
   C.  Both graphs                         D.  Neither graph

6. A graph has eight vertices, and each vertex has degree 4. If the graph is Hamiltonian, how many edges will the Hamilton cycle contain?

   A.  7             B.  8             C.  12           D.  16

7. For the graph below, what is the cost of the Hamilton cycle obtained by using the sorted-edges algorithm?

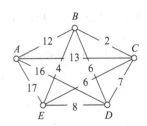

A. 44            B. 45            C. 46            D. 58

8. For the graph in the preceding question, what is the cost of the Hamilton cycle obtained by using the nearest-neighbor algorithm, starting from $A$?

   A. 44            B. 45            C. 46            D. 58

9. Repeat the preceding question, but use the nearest-neighbor algorithm starting from $E$.

   A. 44            B. 45            C. 46            D. 58

10. In which of the following cases would it be helpful to use a solution to the Traveling Salesman problem?
    A.  The city wishes to arrange its garbage collection route
    B.  An airline needs to monitor the check-in procedures at all its airports
    C.  Both A and B
    D.  Neither A nor B

## Exercises 7

1. How many edges are there in the following graphs?

   (i)  $C_6$                              (ii)  $P_5$

2. How many edges are there in the following graphs?

   (i)  $K_5$                              (ii)  $K_{3,2}$

3. How many Hamilton circuits does a 9-vertex complete graph contain?
4. Find a Hamilton cycles in each of the following graphs:

   (i)                                     (ii)

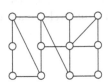                       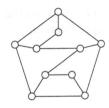

**5.** Find a Hamilton cycles in each of the following graphs:

(i)

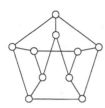

(ii)

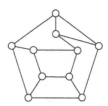

**6.** Find a Hamilton cycles in each of the following graphs:

(i)

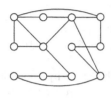

(ii)

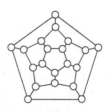

**7.** Prove that the following graph contains no Hamilton cycle.

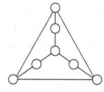

**8.** In each case, does the graph contain a Hamilton path? If so, find one.

**9.** List all Hamilton cycles in the graphs shown.

(i)

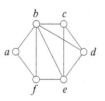

(ii)

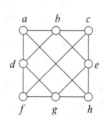

**10.** List all Hamilton cycles in the graphs shown.

(i)  (ii)

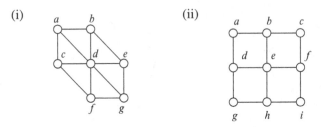

**11.** Consider a complete graph $K_5$ with vertices $abcde$, the costs associated with the edges being

$$ab = 1 \quad ac = 2 \quad ad = 3 \quad ae = 4 \quad bc = 7$$
$$bd = 8 \quad be = 9 \quad cd = 5 \quad ce = 6 \quad de = 10.$$

Find the cost of the cheapest Hamilton cycle in this graph by a complete search. Then find the costs of the routes generated by the nearest-neighbor algorithm starting at each of the five vertices in turn and show that the nearest-neighbor solution is never the cheapest.

In Exercises 12–19 you are given the costs associated with the edges of a complete graph with vertices $abcde$. Find the costs of the routes generated by the nearest-neighbor algorithm starting at each of the five vertices in turn and by the sorted-edges algorithm.

**12.** $ab = 24 \quad ac = 26 \quad ad = 20 \quad ae = 21 \quad bc = 33$
$bd = 29 \quad be = 30 \quad cd = 25 \quad ce = 27 \quad de = 28.$

**13.** $ab = 59 \quad ac = 69 \quad ad = 60 \quad ae = 58 \quad bc = 56$
$bd = 69 \quad be = 54 \quad cd = 58 \quad ce = 66 \quad de = 61.$

**14.** $ab = 16 \quad ac = 24 \quad ad = 30 \quad ae = 48 \quad bc = 27$
$bd = 29 \quad be = 44 \quad cd = 16 \quad ce = 46 \quad de = 51.$

**15.** $ab = 91 \quad ac = 79 \quad ad = 75 \quad ae = 82 \quad bc = 87$
$bd = 64 \quad be = 78 \quad cd = 68 \quad ce = 81 \quad de = 88.$

**16.** $ab = 45 \quad ac = 28 \quad ad = 50 \quad ae = 36 \quad bc = 21$
$bd = 42 \quad be = 34 \quad cd = 44 \quad ce = 39 \quad de = 25.$

**17.** $ab = 11 \quad ac = 15 \quad ad = 13 \quad ae = 18 \quad bc = 20$
$bd = 12 \quad be = 16 \quad cd = 14 \quad ce = 19 \quad de = 17.$

**18.** $ab = 14 \quad ac = 12 \quad ad = 15 \quad ae = 24 \quad bc = 27$
$bd = 29 \quad be = 44 \quad cd = 16 \quad ce = 46 \quad de = 51$

**19.** $ab = 44 \quad ac = 49 \quad ad = 56 \quad ae = 52 \quad bc = 45$
$bd = 54 \quad be = 48 \quad cd = 51 \quad ce = 55 \quad de = 50$

**20.** Construct an example of a graph that contains a Hamilton cycle but does not contain an Euler circuit.

**21.** A Hamilton Path is a path that passes through every vertex in a graph exactly once, but does not return to its starting point. Give an example of a situation where one might want a Hamilton path rather than a Hamilton cycle.

# Chapter 8
# More About Graphs

We shall look at two more applications of graphs, involving spanning trees and graph colorings. We shall conclude with an interesting example that shows there is more information in graph structure than you might think.

## 8.1 Trees

We defined a *tree* to be a connected graph that contains no cycles. Trees occur in a number of applications, and we shall examine some of them here.

We begin with a few examples.

**Sample Problem 8.1** *Which of the following graphs are trees?*

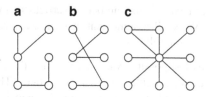

**Solution.** Graph (a) is a tree. Graph (b) is not a tree because it is not connected, although this might not be obvious immediately. Graph (c) contains a cycle, so it is not a tree.

**Your Turn.** Which of the following graphs are trees?

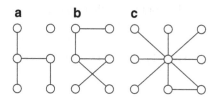

W.D. Wallis, *Mathematics in the Real World*, DOI 10.1007/978-1-4614-8529-2_8,
© Springer Science+Business Media New York 2013

The tree diagrams used in Chap. 3 are just trees laid out in a certain way. To form a tree diagram from a tree, select any vertex as a starting point. Draw edges to all of its neighbors. These are the first generation. Then treat each neighbor as if it were the starting point; draw edges to all its neighbors except the start. Continue in this way; to each vertex, attach all of its neighbors except for those that have already appeared in the diagram. The diagram used in constructing all Hamilton cycles, in Chap. 7, is also a tree diagram.

Every path is a tree. Another special type of tree is called a star; this is a complete bipartite graph with only one vertex in one of its parts, or $K_{1,n}$.

A tree is a minimal connected graph in the following sense: if any vertex of degree at least 2, or any edge, is deleted, then the resulting graph is not connected. In fact it is easy to prove that a connected graph is a tree if and only if every edge has this property.

Trees can also be characterized among connected graphs by their number of edges: a tree has precisely one fewer edge than it has vertices. We shall prove this as a formal theorem.

The proof is an example of the technique called *proof by contradiction*. Suppose you want to prove that a certain statement is true. You start by assuming the statement is *false*. Then you make some deductions using your assumption. Eventually you "prove" something that you know is false—for example, that a certain number is both odd and even, or that $0 = 1$. Something must be wrong with your argument, and the only candidate is your assumption. So the assumption is wrong; the original statement must have been true, not false.

**Theorem 2.** *A finite connected graph G with v vertices is a tree if and only if it has exactly $v - 1$ edges.*

**Proof.**

(i) Let us call a tree *irregular* if its number of vertices is not greater by 1 than its number of edges. We shall assume that at least one irregular tree exists, and show that a contradiction follows.

There must be a smallest possible value $v$ for which there is an irregular tree on $v$ vertices, and let $G$ be an irregular tree with $v$ vertices. This value $v$ must be at least 2, because the only graph with one vertex is $K_1$, which has $v - 1(= 0)$ edges. Choose an edge in $G$ ($G$ must have an edge, or it will simply be a collection of $v$ vertices, with no edges joining them, and will not be connected) and delete it. The result is a union of two disjoint components, each of which is a tree with fewer than $v$ vertices; say the first component has $v_1$ vertices and the second has $v_2$, where $v_1 + v_2 = v$. Neither of these graphs is irregular, so they have $v_1 - 1$ and $v_2 - 1$ edges respectively. Adding one edge for the one that was deleted, we find that the number of edges in $G$ is

$$(v_1 - 1) + (v_2 - 1) + 1 = v - 1.$$

This contradicts the assumption that $G$ was irregular. So there can be no irregular trees; we could not even choose one smallest example. We have shown that a finite connected graph $G$ with $v$ vertices is a tree *only if* it has exactly $v-1$ edges.

(ii) On the other hand, suppose $G$ is connected but is not a tree. Select an edge that is part of a cycle, and delete it. If the resulting graph is not a tree, repeat the process. Eventually the graph remaining will be a tree, and must have $v-1$ edges. So the original graph had more than $v-1$ edges.                           □

Suppose the tree has $v$ vertices. It then has $v-1$ edges. As we saw in Chap. 6, the sum of all degrees of the vertices equals twice the number of edges, or $2(v-1)$. There can be no vertex of degree 0, since the tree is connected and it is not $K_1$; if $v-1$ of the vertices have degree at least 2, then the sum of the degrees is at least $1+2(v-1)$, which is impossible. So there must be two (or more) vertices with degree 1.

An example of a tree with exactly two vertices of degree 1 is the path $P_v$, provided $v > 1$. The star $K_{1,n}$ has $n$ vertices of degree 1.

**Sample Problem 8.2**  *Find all trees with five vertices.*

**Solution.** If a tree has five vertices, then the largest possible degree is 4. Moreover there are 4 edges (by Theorem 1), so the sum of the five degrees is 8. As there are no vertices of degree 0 and at least two vertices of degree 1, the list of degrees must be one of

$$4,1,1,1,1 \quad 3,2,1,1,1 \quad 2,2,2,1,1.$$

In the first case, the only solution is the star $K_{1,4}$. If there is one vertex of degree 3, no two of its neighbors can be adjacent (this would form a cycle), so the fourth edge must join one of those three neighbors to the fifth vertex. The only case with the third degree list is the path $P_5$. These three cases are

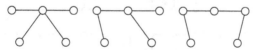

**Your Turn.** Find all trees with four vertices.


## 8.2  Spanning Trees

We shall define a *spanning* subgraph of a graph $G$ to be a subgraph that contains every vertex of $G$. A *spanning tree* in a graph is a spanning subgraph that is a tree when considered as a graph in its own right.

It is easy to show that any connected graph $G$ has a spanning tree. If $G$ is a tree, then the whole of $G$ is itself the spanning tree. Otherwise $G$ must contain a cycle; delete one edge from the cycle. The resulting graph is still a connected subgraph of $G$; and, as no vertex has been deleted, it is a spanning subgraph. Find a cycle in this new graph and delete it; repeat the process. Eventually the remaining graph will contain no cycle, so it is a tree. So when the process stops, we have found a spanning tree.

A given graph might have many different spanning trees. There are algorithms to find all spanning trees in a graph. But fortunately a complete search for spanning trees can be done quite quickly in a small graph. We'll look at an example.

**Sample Problem 8.3** *Find all spanning trees in the following graph.*

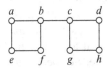

**Solution.** The graph contains two cycles, *abfe* and *cdhg*. In order to construct a tree, it is necessary to delete at least one edge from each of these cycles. As the original graph contains eight vertices, any spanning tree will have eight vertices. From Theorem 2, these trees will have seven edges. So exactly one edge must be removed from each cycle, or there will be too few edges. (This argument would need some modification if an edge that was common to both cycles were deleted, but fortunately the graph contains no such edge.) So there are 16 spanning trees, as follows:

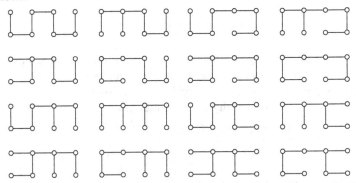

## 8.3   Minimum-Cost Spanning Trees

The following example illustrates an application of spanning trees.

**Sample Problem 8.4** *Suppose an oil field contains five oil wells $w_1$, $w_2$, $w_3$, $w_4$, $w_5$, and a depot $d$. It is required to build pipelines so that oil can be pumped from the wells to the depot. Oil can be relayed from one well to another, at very small cost. All the workers involved will be company employees, so the only real expense is building the pipelines. Figure 8.1a shows which pipelines are feasible*

*to build, represented as a graph in the obvious way, with the cost (in hundreds of
thousands of dollars) shown. (If there is no edge between two vertices, the cost of
a direct join between them would be very high.) Which pipelines should be built?*
**Solution.** Your first impulse might be to connect $d$ to each well directly, as
shown in Fig. 8.1b. This would cost $2,600,000. However, the cheapest solution
is shown in Fig. 8.1c, and costs $1,600,000.

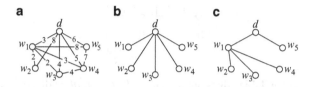

**Fig. 8.1** An oil pipeline problem

Once the pipelines in case (c) have been built, there is no reason to build a direct
connection from $d$ to $w_3$. Any oil being sent from $w_3$ to the depot can be relayed
through $w_1$. Similarly, there would be no point in building a pipeline joining $w_3$ to
$w_4$. The conditions of the problem imply that the graph does not need to contain
any cycle. However, it must be connected. So the solution is to find a spanning tree.
Moreover, the company would prefer the cheapest possible spanning tree. So we are
interested in a *minimum-cost* spanning tree.

We shall look at an algorithm first described by Joseph Kruskal in 1956; not
surprisingly, it is called *Kruskal's algorithm*. Here is how it works.

First of all, select the cheapest edge in the graph. Then select the next cheapest
edge. At every subsequent stage, go through the edges and delete any edge that
would complete a cycle when combined with some of the edges already chosen.
Continue in this fashion. If, at any stage, there are two cheapest edges available,
choose either one. The end result will be a spanning tree, and it can be shown that
this method always gives a cheapest possible tree.

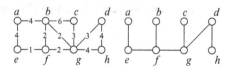

**Fig. 8.2** Finding a minimum-cost spanning tree

As an example, consider the graph on the left of Fig. 8.2. The cheapest edge is
$ef$, cost 1. The next cheapest cost is 2, and we can choose any one of $fg$, $bf$ or $bg$.
Say we select $fg$. For the third edge we can choose either $bf$ or $bg$; whichever one
is chosen, the other will no longer be available, because we need to avoid the cycle
$bfg$. We shall select $fg$. There are two edges of cost 3, namely $cg$ and $dg$. We shall

use *cg*. Then *dg* is still available, so we choose it for the fifth edge. At cost 4, there are four choices: *dg*, *gh*, *ab*, and *ae*. Say we select *dh*; then *gh* must be eliminated. Finally, we choose *ae*, and *ab* is no longer available. The tree is complete, and is shown in the right-hand diagram.

**Sample Problem 8.5** *Suppose you are in the process of using Kruskal's algorithm to find a minimum-cost spanning tree in the graph shown. Heavy line edges are those selected so far. Which edge would you choose next?*

**Solution.** The available edges are *ab*, *cd*, and *fg*. But *ab* is disqualified because it would complete the cycle *abe*. Of the two remaining, *cd* is cheaper, so it is chosen.

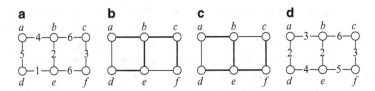

**Fig. 8.3**  An example of Kruskal's algorithm

**Sample Problem 8.6**  *Find a minimal spanning tree in the graph G shown in Fig. 8.3a.*
**Solution.** We start by listing the edges in order of cost: *de*, cost 1; *be*, cost 2; *cf*, cost 3; *ab*, cost 4; *ad*, cost 5; *bc* and *ef*, each cost 6. The first edge chosen is *de*. Next is *be*, then *cf*, then *ab*, cost 4. The next cheapest is *ad*, but it is not allowed because it would form the cycle *abed*. So we go on to either *bc* or *ef*, either may be chosen, and there are two possible solutions, both of total cost 16. They are shown in Fig. 8.3, graphs (b) and (c).
**Your Turn.** Find a minimal spanning tree in the graph *G* shown in Fig. 8.3d.

## 8.4   Representations and Crossings

Consider these two diagrams.

They represent the same graph, the complete graph $K_4$ on four vertices. But as diagrams they are quite different: in the left-hand version, two edges cross; in the right-hand diagram there are no crossings. We shall refer to the two diagrams as different *representations* of $K_4$. The *crossing number* of a representation is the number of different pairs of edges that cross. A graph that can be drawn without any crossings is called *planar* and a drawing of a graph with no crossings is called a *planar representation*. For example, $K_4$ is planar and the right-hand drawing is a planar representation of $K_4$.

**Sample Problem 8.7** *Show that the crossing number of the complete graph $K_5$ on five vertices is* 1.

**Solution.** We can think of $K_5$ as being $K_4$ with one more vertex added. If we start with the representation shown in the left-hand of the diagram, we certainly get at least one crossing, and it is not hard to avoid further crossings. So the crossing number of $K_5$ is either 0 or 1. To obtain a planar representation of $K_5$ we would have to add a new vertex, $e$ say, to the right-hand diagram. But if the new vertex is inside the old diagram, at least one new crossing will be introduced— for example, if it is inside the triangle $bcd$, edge $ae$ must cross one of the original edges—and if $e$ is outside the original diagram, $de$ must cross another edge. The following diagrams illustrate these two cases.

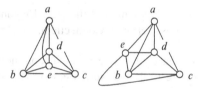

So $K_5$ has crossing number 1.

It is easy to see that any tree has crossing number 0. Similarly, any cycle can be drawn without crossings.

There are many applications of crossing numbers. An early use was in the design of railway yards; it is inconvenient to have the different lines crossing, and sometimes it is preferable to have longer track rather than extra intersections. An obvious extension of this idea is freeway design. At a complex intersection, fewer crossings will mean fewer expensive flyover bridges. More recently, small crossing numbers have proven important in the design of VLSI chips; if two parts of a circuit are not to be connected electrically, but they cross, a costly insulation process is necessary.

## 8.5 Maps and Planarity

By a *map* we shall mean what is usually meant by a map of a continent (showing countries) or a country (showing states or provinces). However we shall make one restriction. Sometimes one state can consist of two disconnected parts (in the United

States, Michigan consists of two separate land masses, unless we consider man-made constructions such as the Mackinaw Bridge). We shall exclude such cases from consideration; in our discussion, any state, or any country, will be a connected area.

Given a map, we can construct a graph as follows: the vertices are the countries or states on the map, and the two vertices are joined by an edge precisely when the corresponding countries have a common borderline. You can draw the graph on top of the map by putting a vertex inside each state and joining vertices by edges that pass through common state borders, so the graph of any map is planar. On the other hand, any planar graph is easily represented by a map.

**Sample Problem 8.8** *Represent the map of the mainland of Australia, divided into states and the Northern Territory, as a graph.*
**Solution.** The map is shown, for reference, alongside its graph.

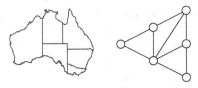

**Your Turn.** Figure 8.4 shows a map of the New England states—Maine, New Hampshire, Massachusetts, Vermont, Connecticut, Rhode Island. Represent it as a graph.

Sometimes two states have only one point in common. An example, in the United States, occurs where Utah and New Mexico meet at exactly one point, as do Colorado and Arizona. We shall say that states with only one point in common have no border, and treat them as if they do not touch.

**Fig. 8.4** New England states

# 8.6 Map and Graph Colorings

In 1852, William Rowan Hamilton wrote to Augustus de Morgan concerning a problem that had been posed by a student, Frederick Guthrie. Guthrie said: cartographers know that any map (our definition) can be colored using four or less colors; is there a mathematical proof? (Guthrie later pointed out that the question had come from his brother, Francis Guthrie.)

Kempe published a purported proof in 1879. It was thought that the matter was over, but in 1890 Heawood pointed out a fallacy in Kempe's proof. Heawood did however repair the proof sufficiently to prove that every planar map can be colored in five colors.

After Heawood's paper appeared, there was renewed interest in what became known as the four-color problem. Because it was easy to state and tantalizingly difficult to prove, it became one of the most celebrated unsolved problems in mathematics. In 1976, Appel and Haken finally proved that any map can be colored in at most four colors. Their proof involved computer analysis of a large number of cases, so many that human analysis of all the cases is not feasible. So we have

**Theorem 3 (The Four-Color Theorem).** *Every map can be colored using at most four colors.*

We define a coloring (or vertex-coloring) of a graph to be a way of applying a set of labels, called *colors*, to the vertices. The coloring is called *proper* or *correct* if no two vertices that receive the same color are neighbors. So instead of coloring a map we could instead apply the colors to the vertices of the corresponding graph. The result is to divide the set of vertices into disjoint subsets, where each subset consists of the vertices that receive a specific color. The subsets are called *color classes*. A *proper coloring* of $G$ is a coloring in which no two adjacent vertices belong to the same color class. In symbols, if $\xi(x)$ denotes the color assigned to vertex $x$, then

$$x \sim y \Rightarrow \xi(x) \neq \xi(y).$$

A proper coloring is called an *n-coloring* if it uses $n$ colors; if $G$ has an $n$-coloring, then $G$ is called *n-colorable*. So the Four-Color Theorem says that every planar graph is four-colorable.

The *chromatic number* $\chi(G)$ of a graph $G$ is the smallest integer $n$ such that $G$ has an $n$-coloring. A coloring of $G$ in $\chi(G)$ colors is called *minimal*. We use the phrase "$G$ is *n-chromatic*" to mean that $\chi(G) = n$ (but note that a minority of authors use *n*-chromatic as a synonym for *n*-colorable).

Incidentally, $\xi$ and $\chi$ are the Greek letters "xi" and "chi." In Greek words, $\xi$ is pronounced like ks in English, while $\chi$ has a ch sound.

A cycle of length $v$ has chromatic number 2 if $v$ is even and 3 if $v$ is odd. The star $K_{1,n}$ has chromatic number 2. Clearly $\chi(G) = 1$ if $G$ has no edges and $\chi(G) \geq 2$ if $G$ has at least one edge. The complete graph on $n$ vertices has chromatic number $n$.

There are two useful results when looking for chromatic numbers. The first is that, if the graph $G$ has a subgraph $H$, then $\chi(G)$ cannot be smaller than $\chi(H)$.

The second, known as Brooks' Theorem, says that if $G$ is any graph other than a complete graph or a cycle of odd length, then the chromatic number of a graph G is never greater than the maximum degree in $G$; however, it can be less.

There is no good algorithm to color a graph in the minimum number of colors. One method, *sequential coloring*, proceeds as follows. First, list the vertices in some order. (The usual method is to arrange the vertices in order of increasing degree.) Apply color 1 to the first vertex in the list. Delete that vertex and all its neighbors, and apply color 1 to the first remaining vertex. Again, delete that vertex and all its neighbors. Continue until the list is empty. Then delete all vertices that received color 1, and apply the same technique to the new list, using color 2. And so on.

The sequential coloring method usually gives a good result, but not necessarily a minimal coloring. For example, consider the 6-cycle $abcdef$. It can be colored in two colors (with color classes $\{ace\}$ and $\{bdf\}$), but if the vertices are listed as $a, d, c, f, b, e$ three colors will be needed.

If you want to find a minimal coloring of a relatively small graph, one useful technique is to find the largest complete subgraph in the graph, and color its vertices with different colors. It is then often possible to color the other vertices by brute force (that is, try all possibilities). For example, in the graph of Fig. 8.5, the vertices $x_1$, $x_5$ and $x_6$ form a $K_3$. Start by coloring them with three different colors. It is then easy to see that the whole graph can be colored in three colors. The color classes are:

$$\{x_1, x_4\}, \quad \{x_2, x_5\}, \quad \{x_3, x_6\}.$$

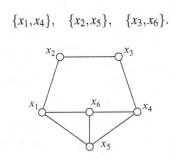

**Fig. 8.5**   A graph with chromatic number 3

Colorings are often used to avoid conflicts. One example is examination scheduling. In this case the vertices represent tests, and an edge means some student must take both the tests represented by the endpoints. So we color the graph; tests receiving same color may be run at the same time. Another is the setting up of a *habitat zoo*. This is a zoo that consists of several large habitats, each containing several species. Obviously one cannot put natural enemies together. So we form a graph whose vertices are species, and an edge means the two species are "enemies." Now color this graph, and put species in same habitat if they receive the same color.

The data for a conflict situation is often given in an array. The rows and columns represent the various elements that may conflict (in our examples, the tests and the species). If elements $i$ and $j$ are incompatible, there is a cross in the $(i, j)$ and $(j, i)$ positions. Replacing crosses by 1s and blanks by 0s produces the incidence matrix of the graph. This information is then used to generate the graph. (Be careful: sometimes authors find it easier to construct an array in which a cross indicates *compatibility*.)

**Sample Problem 8.9** *The following table shows the pairs of species in a habitat zoo that are incompatible.*

| | A | B | C | D | E | F | G | H | I | J |
|---|---|---|---|---|---|---|---|---|---|---|
| A | | | × | | | | | × | × | × |
| B | | | × | × | | | | × | | |
| C | × | × | | | | | | × | × | × |
| D | | × | × | | | | | × | | |
| E | | | | | | × | × | | | × |
| F | | | | | × | | | | | × |
| G | | | | | × | | | | × | × |
| H | × | × | × | × | | | | | × | × |
| I | × | | × | | | | × | × | | × |
| J | × | | × | | × | × | × | × | × | |

*Find the minimum number of habitats that will be needed.*
**Solution.** For convenience, we have drawn the corresponding graph. We need to color it with the minimum number of colors.

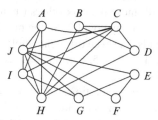

First observe that $A$, $C$, $H$, $I$, and $J$ are all adjacent to each other; they form a complete graph of size 5. So at least 5 colors are needed. Let's use 1 for $A$, 2 for $C$, 3 for $H$, 4 for $I$, and 5 for $J$.

Now look at $B$. It is joined to $C$, $D$, and $H$. So colors 2 and 3 are not available. We shall try color 1 (which is possible because $A$ and B are not joined).

Next consider $D$. It is joined to $B$, $C$, $H$ (and no others). It can take color 4 or 5. We shall try 4.

Finally. a little experimentation shows that $E$, $F$ and $G$ can take colors 1, 2 and 3 respectively with no problem. This is not the only solution, but it shows that five colors is a possibility. Since at least five are needed, the chromatic number is

5 and at least five habitats are required. In our solution, the habitats are $\{A, B, E\}$, $\{C, F\}$, $\{G, H\}$, $\{D, I\}$, and $\{J\}$,

A minimal coloring is not always the best solution to a problem. Sometimes it would be best if the color classes would be about the same size, Such a coloring is called *equitable*. For example, in the habitat zoo problem above, if $B$ is allocated color 5, the five color classes would each have two members. Of course, every graph has an equitable coloring—give every vertex a different color—but typically there will be a limit on the number of colors.

## 8.7   The Handshake Problem

We conclude with an old puzzle that can be solved using a graph representation.

Eight people—four married couples—attend a dinner party. Various introductions are made; everybody shakes hands (once!) with some of the others. Of course, no one shakes hands with his/her spouse.

After a while, the hostess asks every other person in the room, "how many times did you shake hands tonight?" It turns out that no two of the people she asked said that they shook hands the same number of times. (The hostess isn't included of course; she didn't ask herself.)

How many times had the host (the hostess' spouse) shaken hands?

We use graph theory to represent the elements of the problem. Write $W, X, Y, Z$, $w, x, y, z$ for the eight people; $W$ and $w$ are one couple, $X$ and $x$ another, and so on. We represent the party by a graph with people as vertices and handshakes as edges. The number of handshakes is the degree of the vertex.

No one shook hands with more than six others (there are only six other people other than the two members of any couple). So the numbers of times people (other than the hostess) shook hands are seven different whole numbers, ranging from 0 to 6. It follows that each of $0, 1, 2, 3, 4, 5, 6$ occurs exactly once among the degrees of those other than the hostess.

So somebody shook hands six times. Let's call this person $X$. He or she did not shake hands with his/her spouse $x$, so he/she shook with everybody else. These handshakes contribute the edges shown in the figure shown right: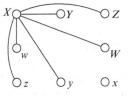

Somebody shook hands 0 times. Looking at the figure, every vertex received at least one edge, except for $x$. (There might be further edges, not yet shown.) So $x$ must be the person with no edges (no handshakes). No matter what happens, you will never add an edge touching $x$.

Now, somebody shook hands exactly five times. It was neither $X$ nor $x$. Suppose it was $Y$. There is already one edge touching $Y$, the edge

*XY*, so there are exactly four more edges touching *Y*. None of the new edges touch *x* or *y* (remember, people don't shake with spouses). This implies that *y* is the only person to receive exactly only one: *x* has 0, and every other vertex has at least 2. The figure looks like the following (possibly with more edges):

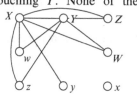

Next, someone shook hands exactly four times. It cannot be *X*, *x*, *Y* or *y*; say it is *Z*. There are already two edges touching *Z*, and *Z* is not joined to *z*. So *Z* must be joined to *W* and *w*. We have obtained the following graph, showing all the handshakes:

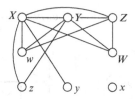

In this graph there are two vertices of degree 3 (*W* and *w*), and each of 0, 1, 2, 3, 4, 5 occurs as a degree once. One of the two vertices of degree 3 must be the hostess (this is the only possible way to get a repeat), and the other is her spouse. So the host shook hands three times.

# Multiple Choice Questions 8

**1.** Which of the following graphs are trees?

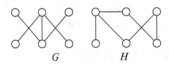

G          H

    A.   *G* only                    B.   *H* only
    C.   Both *G* and *H*          D.   Neither *G* nor *H*

**2.** In which of the diagrams below do the heavy edges represent spanning trees?

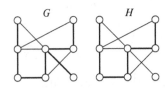

G          H

    A.   *G* only                    B.   *H* only
    C.   Both *G* and *H*          D.   Neither *G* nor *H*

**3.** A graph *G* has 12 vertices, each of degree 4. *T* is a spanning tree in *G*. How many edges does *T* contain?

    A.  11            B.  12           C.  18          D.  24

**4.** When Kruskal's algorithm is used to find a minimum-cost spanning tree on a graph, which of the following is necessarily *false*?

    A.   The tree contains the edge of the graph of minimum cost
    B.   The tree is not necessarily connected
    C.   The tree may contain the edge of the highest cost
    D.   Both B and C are necessarily false

**5.** Which of the following statements is true?

    A.   Every graph has a unique spanning tree
    B.   Every connected graph has a unique spanning tree
    C.   Every connected graph has a unique minimum-cost spanning tree
    D.   None of the above

**6.** Which of the following graph labelings is a proper vertex coloring? (The colors are called $1, 2, 3 \ldots$)

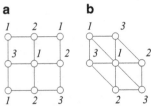

    A. Both graphs (a) and (b)     B. Graph (a) only

    C. Graph (b) only            D. Neither graph (a) nor (b)

7. Which of the following describes what is required of a proper vertex coloring?

    A. All available colors are used

    B. The minimum number of available colors are used

    C. Every edge connects vertices of the same color

    D. No edge connects vertices of the same color

8. Consider the set of all graphs that contain a cycle of length 5. Which of the following is true?

    A. Some graphs have chromatic number 2, the others have chromatic number 3

    B. All the graphs have chromatic number 3

    C. All the graphs have chromatic number at least 3

    D. All the graphs have chromatic number at least 5

9. Which statements is true of graph colorings?

    A. All proper colorings are minimal

    B. All minimal colorings are proper

    C. Both A and B are true

    D. Neither A nor B is true

10. An architect needs to design an intercom system for a large office building. Which of the following techniques is most likely to be useful in solving this problem?

    A. Finding an Euler circuit

    B. Applying an algorithm for the traveling salesman problem

    C. Finding a minimum-cost spanning tree

    D. Construct a minimal vertex coloring

11. A company needs to dispose of quantities of five chemical compounds, $A,B,C,D,E$. A cross in the table indicates that mixing the two compounds causes a dangerous reaction. Which graph would be used to work out which compounds can safely be disposed of together?

|   | A | B | C | D | E |
|---|---|---|---|---|---|
| A |   | × |   |   | × |
| B | × |   | × |   | × |
| C |   | × |   | × |   |
| D |   |   | × |   |   |
| E | × | × |   |   |   |

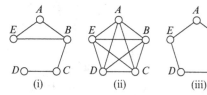

(i)         (ii)         (iii)

    A. (i).       B. (ii).       C. (iii).       D. none of them.

12. The city needs to inspect the roads after a storm, to remove tree branches. Which of the following will be most helpful to them?

    A.  Construct a minimal Eulerization.
    B.  Construct a minimum-cost Hamilton circuit.
    C.  Construct a minimum-cost spanning tree.
    D.  Construct a minimal vertex coloring.

# Exercises 8

1. A tree has eight vertices. How many edges does it contain?
2. Suppose a graph has six vertices.
   (i) Suppose two vertices are of degree 3 and four of degree 1. Are the vertices of degree 3 adjacent, or not, or is it impossible to tell?
   (ii) Two vertices are of degree 4 and four of degree 1. Is such a graph possible?
3. Find all trees on six vertices, in which precisely four vertices have degree 1.
4. Find all trees with seven vertices, precisely three of them of degree 1.
5. Find spanning trees in the following graphs.

       (i)          (ii)        (iii)      (iv)

*In Exercises 6–9, find the number of spanning trees in the graph, and sketch all the trees.*

6.

7.

8.

9.

*In Exercises 10–19, find minimal spanning trees in the graph.*

10.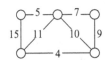

11. 

12.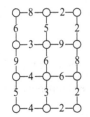

13.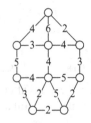

**14.** 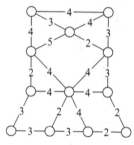     **15.**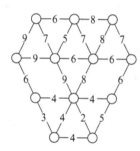

**16.** 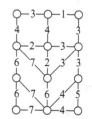     **17.**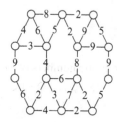

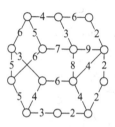     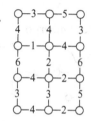

**18.** 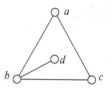     **19.**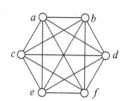

**20.** The graph $G$ is obtained from $K_5$ by deleting one edge. Show that $G$ can be colored with four colors.

**21.** $G$ is formed from the complete graph $K_v$ by deleting one edge. Prove that $\chi(G) = v - 1$ and describe a way of coloring $G$ in $v - 1$ colors.

In Exercises 22–27, is the graph planar? What is the chromatic number? Find a minimal coloring.

**22.**

**23.**

**24.** 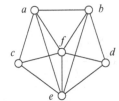         **25.**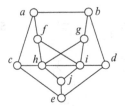

**26.** 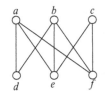         **27.**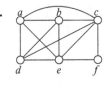

**28.** What are the chromatic numbers of the graphs $A, B$, and $C$?

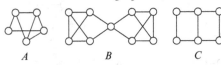

                        $A$                  $B$                $C$

**29.** What are the chromatic numbers of the graphs $D, E$, and $F$?

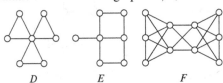

                $D$        $E$            $F$

In Exercises 30–33, a company executive needs to schedule a number of committee meetings in the minimum number of time slots. In the array, a cross shows that the two committees have a common member, so they cannot meet at the same time. What is the minimum number of time slots needed? Sketch the relevant graph.

**30.**

|   | A | B | C | D | E | F |
|---|---|---|---|---|---|---|
| A |   | × |   |   |   | × |
| B | × |   | × |   |   | × |
| C |   | × |   | × | × |   |
| D |   |   | × |   | × |   |
| E |   |   | × | × |   |   |
| F | × | × |   |   |   |   |

**31.**

|   | A | B | C | D | E | F |
|---|---|---|---|---|---|---|
| A |   | × | × |   |   | × |
| B | × |   |   | × |   |   |
| C | × | × |   | × |   |   |
| D |   |   | × |   | × | × |
| E | × |   |   | × |   | × |
| F | × |   |   |   | × | × |

**32.**

| | A | B | C | D | E | F | G | H |
|---|---|---|---|---|---|---|---|---|
| A | | × | | | | | × | |
| B | × | | × | | | | | × |
| C | | × | | × | | | | |
| D | | | × | | × | | | × |
| E | | | | × | | × | | |
| F | | | | | × | | | × |
| G | × | | | | | | | × |
| H | | × | | × | | × | × | |

**33.**

| | A | B | C | D | E | F | G | H |
|---|---|---|---|---|---|---|---|---|
| A | | × | | | | | × | |
| B | × | | × | × | | | × | × |
| C | | × | | × | | | | |
| D | | × | × | | × | | | × |
| E | | | | × | | × | | × |
| F | | | | | × | | × | × |
| G | × | × | | | | × | | × |
| H | | × | | × | × | × | × | |

**34.** In Exercise 33, one of the members of Committee *H* wishes to join one of the other committees in addition. If no further time slots are available, which committees can she join?

**35.** (i) Show that the chromatic number of the graph shown is 4.
    (ii) Find an edge in the graph such that, if it is deleted, the resulting graph will have chromatic number 3.

**36.** The illustration shows a two-coloring of a graph. Find an ordering of the vertices such that the sequential coloring algorithm will require four colors.

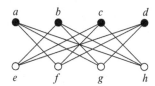

# Chapter 9
# Identification Numbers

## 9.1 Simple Check Digits

In this chapter we look at some examples of how identification numbers such as Social Security Numbers, Zip Codes and Bank Account Numbers are made up, and how and why they are used. We see these numbers all the time. For example, when you go to a supermarket, the string of black lines printed on a packet, called a *barcode*, represents a string of numbers. The cash register scans the bars and converts them into a string of numbers; the corresponding prices are kept in a file and are looked up automatically.

Many other things in everyday life are also represented by codewords—strings of letters or numbers which identify a bank account, a book, an airline ticket, and so on. When these are copied or transmitted there is always the possibility of errors.

The simplest method to avoid errors is a check digit. For example, a Visa or American Express traveler's check has a 10-digit identifying code, such as 2411903043. The first nine digits identify the check; the tenth is chosen so that the sum of all ten digits is divisible by 9. In the example, $2+4+1+1+9+0+3+0+4+3 = 27 = 3 \times 9$. If the first nine digits were 524135324, we observe that $5+2+4+1+3+5+3+2+4 = 29$; the next multiple of 9 is $29+7 = 36 = 4 \times 9$, so the check digit would have to be 7, and the code is 5241353247. If the check sum is wrong, the number was wrong. This simple method detects most errors consisting of one wrong symbol (but not when a 0 is substituted for a 9 or vice versa; in some systems of this sort, no 9s are ever used). Of course, if you get two digits wrong, the check will not always work, and it never helps when you interchange two digits.

A modification of this method is used in US Postal Service money orders. The money order has an 11-digit number; the first ten digits identify the money order, and the 11th is a check digit. Again, divisors on division by 9 are involved. But in this case, after the sum of the first ten digits is found and divided by 9, the check digit equals the remainder. For example, if the identification number started 2411403043, then $2+4+1+1+4+0+3+0+4+3 = 22 = 2 \times 9+4$, so the last digit would be 4.

W.D. Wallis, *Mathematics in the Real World*, DOI 10.1007/978-1-4614-8529-2_9,
© Springer Science+Business Media New York 2013

**Sample Problem 9.1** *A USPS money order has number 5_2413532404, where the second digit has been rubbed off. What was the second digit?*
**Solution.** Represent the second digit by $x$. The sum of the first ten digits is $29 + x$. When this is divided by 9, the remainder must be 4, so $29 + x$ is one of $13, 22, 31, 40, \ldots$ But $0 \le x \le 9$, so the only possibility is 31, and $x$ must equal 2.
**Your Turn.** What is the missing digit in the USPS money order with identification 31214_25255?

These systems will detect some errors, but by no means all. For example, one of the most common errors we make is to write digits in the wrong order, especially to exchange two adjacent numbers. If you were to swap two digits on a traveler's check number, the error would not be noticed. The money order number system is slightly better; if you were to exchange the *last* digit with another, the error would probably be caught. But these methods do not detect all errors by any means.

## 9.2   Codabar

The Codabar system is used by many credit card companies, and also by libraries, blood banks, and other companies.

A Codabar number has an even number of digits: an odd number of digits to identify the account, plus one check digit. There are five steps in calculating the check digit:

1. add together the digits in the *odd-numbered* positions (first, third, and so on) of the identity number;
2. multiply this sum by 2;
3. count how many of the odd-numbered digits are greater than 4, and add the answer to the product from step 2;
4. add the digits in the even-numbered positions of the identity number, and add this sum to the result of step 3;
5. the check digit is the number that, when added to the result of step 4, would make the total exactly divisible by 10.

**Sample Problem 9.2** *What is the check digit that would be attached to the Codabar identity number* 3125600196431?
**Solution.** The odd-position digits are 3,2,6,0,9,4,1, and their sum is 25. So the answer to step 1 is 25 and, to step 2, 50. Two of these digits are greater than 4, so the answer to step 3 is 52. The even-position digits, 1,5,0,1,6,3, have sum 16, giving total 68. So the check digit is 2.

To check whether a number is a legitimate Codabar number, we ignore the last digit and calculate the check digit for the remaining number.

**Sample Problem 9.3** *Is* 3125700143750015 *a legitimate Codabar number?*
**Solution.** $3+2+7+0+4+7+0+1 = 24$; twice this is 48. There are two digits over 4, so the total becomes 50. The even digits add to 15, giving a total of 65. So the check digit is 5, and the number is legitimate.
**Your Turn.** Is 55914308123335 a legitimate Codabar number?

The Codabar method detects a lot more errors than the simpler methods we saw in the preceding question. In particular, if two adjacent digits are exchanged, the odd-position and even-position sums are changed, and in most cases the check digit will be altered. It has been estimated that Codabars catch about 98% of errors.

## 9.3 ISBNs

The International Standard Book Numbering system is an interesting example of a code that detects two sorts of errors: a single mistaken digit or a transposition of two digits. The ISBN is a 10-digit code representing a book. Say the code for a particular book is $ABCDEFGHIJ$. Then:

- the first digit indicates the language in which the book is published (not necessarily the language in which the book is written: for example, Spanish textbooks published in the United States have an ISBN starting with 0, meaning English, even if the book is an immersion textbook, written entirely in Spanish);
- the next three or four digits are a code for the publisher;
- the next few digits—all but the last—indicate the particular book;
- the last digit, $J$, is chosen so that

$$10A + 9B + 8C + 7D + 6E + 5F + 4G + 3H + 2I + J$$

is divisible by 11.

There is a possible problem here. Suppose $10A + 9B + 8C + 7D + 6E + 5F + 4G + 3H + 2I$ leaves a remainder 1 on division by 11. It is necessary to have $J = 10$, and there is no single digit available. The standard remedy is to write $X$ for the last digit in this case.

Suppose a digit is written incorrectly in an ISBN. For example, suppose that instead of $ABCDEFGHIJ$, one writes $ABCdEFGHIJ$, where $D$ and $d$ are different. Then when you test for divisibility by 11, the check sum

$$10A + 9B + 8C + 7d + 6E + 5F + 4G + 3H + 2I + J$$

is calculated. Now

$$10A + 9B + 8C + 7d + 6E + 5F + 4G + 3H + 2I + J$$
$$= 10A + 9B + 8C + 7D + 6E + 5F + 4G + 3H + 2I + J + 7(d - D),$$

and as $d \neq D$ this difference $7(d - D)$ is a product of two numbers both less than 11 in size. So it is not divisible by 11. So the supposed check sum is not divisible by 11. The same argument applies if any one digit is written incorrectly.

On the other hand, suppose the correct numbers are written, but two of them are transposed. The argument does not depend on which pair; as an example, suppose the $C$ and $H$ are interchanged. Then the check sum is

$$10A + 9B + 8H + 7D + 6E + 5F + 4G + 3C + 2I + J$$
$$= 10A + 9B + 8C + 7D + 6E + 5F + 4G + 3H + 2I + J + 5(H - C);$$

and again the check sum is wrong, because $H - C$ is less than 11.

The standard way of writing an ISBN is with dashes after the first, fifth and ninth digit: for example, 0-8176-8319-$X$.

**Sample Problem 9.4** *Find the check digit for an ISBN number that starts* 0-6693-3907.

**Solution.** Multiplying, $10 \times 0 + 9 \times 6 + 8 \times 6 + 7 \times 9 + 6 \times 3 + 5 \times 3 + 4 \times 9 + 3 \times 0 + 2 \times 7 = 0 + 54 + 48 + 63 + 18 + 15 + 36 + 0 + 14 = 248$. Now $248 = 242 + 6 = 11 \times 22 + 6$, So we must add 5 to obtain a multiple of 11, and the check digit is 5.

**Your Turn.** Verify that the number 0-8176-8319-$X$, given above is a legitimate ISBN number.

In 2007, a new version of the ISBN was introduced. It has 13 digits, so it is referred to as the ISBN-13, and the original is now called the ISBN-10. The ISBN starts with either the three digits 978 or 979. The next nine digits are the same as the first nine digits of the ISBN-10. Finally, the check digit is appended. It is defined as follows: the sum of the even-position digits plus three times the sum of the odd-position digits must be divisible by 10. The use of 10, rather than 11, eliminates the need for the symbol $X$ as a possible check digit, but more errors may pass undetected.

The ISBN-13s are a class of *International Article Numbers*, also called EANs, 13-digit codes that are allocated to a large number of different retail articles. (They were originally called European Article Numbers, whence the abbreviation.) The introductory code numbers 978 and 979 are allocated to book manufacturers, and other introductory codes are assigned to other products. These numbers are used in making the barcodes that are scanned when items are purchased.

**Sample Problem 9.5** *What is the ISBN-13 corresponding to the ISBN 0-8176-8319-X, assuming that the first three digits are 978?*

**Solution.** The first 12 digits are 978081768319. Suppose the check digit is $x$. Then the sum of the odd positions is $40 + x$ and the sum of the even positions is 27. So $3(40 + x) + 27 = 147 + 3x$ must be a multiple of 10. The only single-digit possibility is $x = 1$. The ISBN-13 is 978-0-8176-8319-1.

## 9.4   The Soundex

The *soundex* system to encode names was developed in 1918 by Robert C. Russell. It gives a short string to show roughly how the first part of a name (or other word) sounds when spoken. To find the soundex of a name, first write the letters of the name. Then go through the following steps:

1. delete the first letter of the name (but keep a record of it);
2. delete any A, E, I, O, U, Y, H, or W;
3. replace letters by numbers, as follows:
   use 1 for B, F, P or V;
   use 2 for C, G, J, K, Q, S, X or Z;
   use 3 for D or T;
   use 4 for L;
   use 5 for M or N;
   use 6 for R;
4. if two or more adjacent numbers are the same, omit all but one;
5. if there are more than three numbers, keep only the first three; if fewer than three, add 0s to the end to make the code up to three numbers;
6. put the first letter of the original name on the front of the code.

**Sample Problem 9.6** *Find the soundex representations of the names SPRINGER and HAMMAN.*
**Solution.**

|     | SPRINGER | HAMMAN |
| --- | --- | --- |
| 1.  | PRINGER  | AMMAN  |
| 2.  | PRNGR    | MMN    |
| 3.  | 16526    | 555    |
| 4.  | 16526    | 5      |
| 5.  | 165      | 500    |
| 6.  | S165     | H500   |

## 9.5  Driver's Licenses

The soundex is used in constructing driver's licenses in a number of states. We shall go through the Illinois and Florida license number constructions.

The Illinois license number consists of one letter followed by 11 digits. We shall work through the construction of the number for Carl D. Springer, born on May 4, 1977.

The first four symbols are the soundex of the license-holder's last name. As we saw above, Springer has soundex S165.

The next three characters are formed as follows: an arbitrary 3-digit number is assigned for your first name—for example,

140 for Charles, Clara
120 for Carl, Catherine
100 for all other names beginning with C

and another number (from 1 to 19) is added for your middle initial, if any (1 for A, 6 for F, 18 for T, U or V, and so on). A list of name encodings is shown in the table below. Carl is 120 and D is 4, so Mr Springer's 5th, 6th and 7th digits are 124.

The last five are formed from your sex and date of birth:

ABCDE means the driver was born in the year 19AB; the CDE is derived from his or her birthday. First you calculate the day of the year in which the applicant was born—approximately. For some reason, the calculation is performed as though every month had 31 days. For example, if the birthday was April 2nd, you work as though January, February and March each had 31 days, so April 2nd would be the 95th day of the year ($95 = 3 \times 31 + 2$), even though it is actually the 92nd or 93rd, depending on whether it is a leap year. So, if the driver is a man, you would put CDE = 095. For a woman, increase the total by 600; a woman born on April 2nd has CDE = 695. For Mr Springer, May 4 is 128, so his last five digits are 77128, and his license number is S165-1247-7128.

**Sample Problem 9.7** *What would be the Illinois license number for Catherine A. Churchill, born May 2, 1976?*
**Solution.** Churchill: C624
Catherine: 120
Middle initial A: add 1, gives 121
Birth year: 76
Birth date: $4.31 + 2 = 126$
Woman: add 600, gives 726
                    C624-1217-6726.
**Your Turn.** What would be the Illinois license number for Anne E. Hamman, born August 10, 1988?

# Illinois Driver's License Tables

Here are the standard numbers for some common first names.

| | | | |
|---|---|---|---|
| 020 | Albert, Alice | 040 | Ann, Anna, Anne, Annie, Arthur |
| 080 | Bernard, Bette, Bettie, Betty | 120 | Carl, Catherine |
| 140 | Charles, Clara | 180 | Dorothy, Donald |
| 220 | Edward, Elizabeth | 260 | Florence, Frank |
| 300 | George, Grace | 340 | Harold, Harriet |
| 360 | Harry, Hazel | 380 | Helen, Henry |
| 440 | James, Jane, Jayne | 460 | Jean, John |
| 480 | Joan, Joseph | 560 | Margaret, Martin |
| 580 | Marvin, Mary | 600 | Melvin, Mildred |
| 680 | Patricia, Paul | 740 | Richard, Ruby |
| 760 | Robert, Ruth | 820 | Thelma, Thomas |
| 900 | Walter, Wanda | 920 | William, Wilma |

Some minor variants may be given numbers from this list—for example, I would guess Will would be classified as 920. However, if your name is not listed, it will usually be given a code corresponding to its first letter:

| | | | | | | | |
|---|---|---|---|---|---|---|---|
| A | 000 | H | 320 | O | 640 | V | 860 |
| B | 060 | I | 400 | P | 660 | W | 880 |
| C | 100 | J | 420 | Q | 700 | X | 940 |
| D | 160 | K | 500 | R | 720 | Y | 960 |
| E | 200 | L | 520 | S | 780 | Z | 980 |
| F | 240 | M | 540 | T | 800 | | |
| G | 280 | N | 620 | U | 840 | | |

For example, Eric should be 200. If you use only the first initial of your first name, this table is used. If your license just says J. Smith you get 420. Even if your name is John Smith, it is 420, not 460—they go by the name as printed on the license.

For your middle initial, use

| | | | | | | | |
|---|---|---|---|---|---|---|---|
| A | 1 | H | 8 | O | 14 | V | 18 |
| B | 2 | I | 9 | P | 15 | W | 19 |
| C | 3 | J | 10 | Q | 15 | X | 19 |
| D | 4 | K | 11 | R | 16 | Y | 19 |
| E | 5 | L | 12 | S | 17 | Z | 19 |
| F | 6 | M | 13 | T | 18 | | |
| G | 7 | N | 14 | U | 18 | | |

It is quite possible for two people to get the same number. In that case an *overflow* number is appended. In Illinois these are not printed on the license, but the police have access to it.

Illinois also issues a state ID. It is often used as a proof of age or identity by people who do not drive, and is also a good idea to have one when you are traveling, in case your license is misplaced or stolen. The ID has a number derived in the same way, except that the letter is put at the end. Mr Springer's ID number would be 1651-2477-128S, and Ms Churchill would get ID number 6241-2176-726C.

The Florida driver's license is very similar to the Illinois one, with three differences. First, the day-of-birth calculation is done as if there are 40 days in each month. Second, the amount added to a woman's day of birth is 500, not 600. And finally, the overflow number is printed on the license, a single digit at the end of the number. (If there were no duplicates when the license was issued, the number is 0.)

The spacing is a little different: assuming Mr Springer applied for a Florida license and two other people had already come up with the same number, so that his overflow number was 2, his license number would be S165-124-77-128-2.

**Sample Problem 9.8** *What would be the Florida license number for Catherine A. Churchill, born May 2, 1976, assuming there is no overflow?*
**Solution.** C624-121-76-662-0.
**Your Turn.** What would be the Florida license number for Anne E. Hamman, born August 10, 1988, if there is no overflow?

# Multiple Choice Questions 9

1. Determine the check digit that should be appended to the US Postal Service money order identification number 0849202911.

    A.  1        B.  4        C.  5        D.  8

2. Suppose a US Postal Service money order is numbered ?4839203210 where the first digit is obliterated. What is the missing digit?

    A.  4        B.  5        C.  8

    D.  The missing digit can't be determined

3. A Visa traveler's check has identity number 314247605. What is the check digit?

    A.  2        B.  4        C.  5        D.  8

4. What is the check digit for the Codabar identity number 400521600193361?

    A.  2        B.  4        C.  8        D.  9

5. Suppose the first digit of the ISBN-10 ?-7127-1011-0 is obliterated. The question mark represents the missing digit. What is it?

    A.  5        B.  6        C.  9        D.  0

6. What is the check digit for the ISBN-10 0-1011-2102-?

    A.  4        B.  2        C.  $X$        D.  7

7. Determine the check digit that should be appended to the ISBN-10 0-7167-9811.

    A.  2        B.  5        C.  6        D.  9

8. What is the check digit for the ISBN-13 978-0-1011-2102-?

    A.  4        B.  2        C.  $X$        D.  7

9. A woman's birthday is February 5, 1987. What are the last four digits on her Illinois driver's license?

    A.  7636        B.  7036        C.  3287        D.  2387

10. Suppose a man born on February 1st, 1977 receives an Illinois driver's license. How would the date of his birth be encoded on the license?

    A.  77532        B.  M20177        C.  77632        D.  77032

# Exercises 9

1. Determine the check digit for the Visa traveler's check number that starts 241198304.
2. Determine the check digit that should be appended to American Express Travelers Cheque identification number 293021243.
3. If the third digit of the Visa traveler's check number 22387235443 is mistyped, can the check digit detect the error? Explain.
4. What is the check digit for a US Postal Service money order if the number starts 1144326067?
5. Suppose a US Postal Service money order is numbered $x$3843591010 where $x$ means that the first digit is obliterated. What is the missing digit?
6. Explain why the Postal Service money order scheme cannot detect the mistake of substituting a 9 for a 0.
7. Use the Codabar scheme to determine the check digit for the number:
   2395608197451
8. Is 54932746437963 a valid number in the Codabar system?
9. A 10-digit ISBN number starts 0-6680-9325. What should the check digit be?
10. The fifth digit of an ISBN-10 number has been mistyped; the number appears as 0-1134-7251-3. What should it be?
11. Is 0-5342-9764-3 a valid ISBN-10 number?
12. If the first and second digits of the ISBN-10 number 0-7167-2378-6 are exchanged, can the system detect an error? Can it recover the original number? Explain.
13. What is the Soundex representation for the name Blackwell?
14. What is the Soundex representation for the name Dicksson?
15. What is the Soundex representation for the name Simpson?
16. What is the Soundex representation for the name Warren?
17. A woman called Margaret M. Morrison, born April 3rd, 1992, receives an Illinois driver's license. What is the license number? If she received a state ID, what would be the number?
18. A man called John T. Perkins was born on May 4th, 1991, and receives a Florida driver's license. No one with the same birth date and surname soundex has previously received a Florida Driver's license. What is his license number?
19. A woman called Wanda R. Bloomfield, born December 3rd, 1988, receives an Illinois driver's license. What is the license number? If she received a Florida license, and one other person has already received a number with the same ID, what would be the number?
20. A man called Edward J. Berry, born May 1st, 1966, receives an Illinois driver's license. What is the license number? If he received a Florida license, and no other person has already received a number with the same ID, what would be the number?

# Chapter 10
# Transmitting Data

## 10.1 Errors

Errors occur whenever we transmit data. For example, suppose you are reading a newspaper (in English). If you see the word "bive," you know it is not an English word, so a typographical error has occurred. You can usually correct it by looking at the context.

For example, if you see:

| | | |
|---|---|---|
| "...the next bive years..." | it should say | five |
| "...a bive performance..." | it should say | live |
| "...teeth used in the bive..." | it should say | bite |
| "...bive you an example..." | it should say | give |

Nowadays a lot of data is sent electronically, for example by computer, and is a *binary* message. We usually represent the message by a string of the symbols 0 and 1; these may, for example, represent power on/power off or switch up/switch down. Data on computers or on tools governed by computers is stored in binary form. Errors can be caused by many things (electrical interference, human error, and so on). We need to look at ways to detect and possibly correct these errors. As there is no context, we have a problem. To get around this, a message can be given its own context by sending some more information. We have already seen examples of adding additional information in the preceding chapter, with the check digits appended to ISBNs and other identity strings.

## 10.2 Venn Diagram Encoding

For example, suppose we have a system where the messages consist of four binary digits, such as 1100 or 0111. In this example, to send 110011010110, you might first break it up into

W.D. Wallis, *Mathematics in the Real World*, DOI 10.1007/978-1-4614-8529-2_10,
© Springer Science+Business Media New York 2013

$$1100 \quad 1101 \quad 0110$$

(so you can send it as three 4-symbol messages). Then each set of 4 digits is replaced by a set of seven digits according to a certain plan. This is called *encoding* the message. We shall outline one such scheme, where each four-digit string is replaced by a seven-digit string. The three additional digits are calculated using a diagram like a Venn diagram, so the method is called *Venn diagram encoding*.

We shall represent the original message by the four-digit string ABCD. Each of A, B, C or D represents a 1 or a 0; in the example 1100, A would represent 1, B is 1, C is 0, and D is 0. We transmit a string ABCDEFG.

The set of seven digits is constructed as follows:

First, draw a diagram of three overlapping circles, and write the numbers A, B, C and D in the positions shown in the left-hand diagram in Fig. 10.1. Then write in three more numbers—E, F, G—chosen as follows. In each circle the sum of the four numbers is to be even.

A + B + C + E must be even.

A + C + D + F must be even.

B + C + D + G must be even.

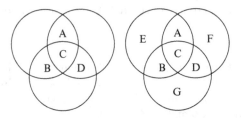

**Fig. 10.1**   Venn diagram encoding

**Sample Problem 10.1** *Encode the message words* 1101 *and* 0110 *using Venn diagram encoding.*

**Solution.** Start by forming the following diagrams:

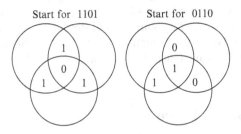

In the diagram for 1101, all three circles contain two 1s, so the sums are all even. Each of the other three symbols will be 0. In the 0110 diagram, the top left circle is even, so it receives 0 for its new symbol; but each of the others has sum 1, which is odd, so those circles receive another 1. The final diagrams are

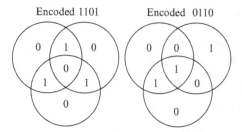

and the strings transmitted will be 1101000 and 0110010 respectively.

**Your Turn.** Encode the message words 1001 and 1111 using Venn diagram encoding.

The number *ABCD* is called the *message word*, and ABCDEFG is called the *codeword*. This method of constructing codewords from strings of length 4 is called *Venn diagram encoding*. The person who receives the message fills the numbers into a diagram and checks the three circle sums:

- If they are all even, it is assumed that the correct message was received.
- If any are odd, corrections are made on the assumption that exactly one number is wrong in each odd circle.

This technique has an obvious flaw: if two or more errors are made, the correct message will not be recovered; in fact, the message will be "corrected" to something wrong. But it is reasonably reliable provided the chance of an error is small, and there are other methods available when errors are more likely.

Suppose you receive the codeword 1010010. When you form the diagram, shown on the left of Fig. 10.2, you see that the sum of the top left circle is even (2), but the top right and bottom sums are both odd (3 and 1 respectively). You can change at most one digit, so the only solution is to change the entry common to the two circles, as shown in the right-hand diagram. The corrected codeword is 1011010, and you believe the intended message was 1011.

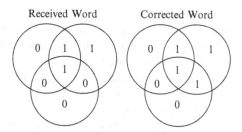

**Fig. 10.2** Correcting a codeword with one error

Suppose the sums of the entries in the top left, top right, and bottom circles are $X, Y$, and $Z$ respectively. Then:

- if only $X$ is odd, change entry E;
- if only $Y$ is odd, change entry F;
- if only $Z$ is odd, change entry G;
- if $X$ and $Y$ are odd, change entry A;
- if $X$ and $Z$ are odd, change entry B;
- if $Y$ and $Z$ are odd, change entry D;
- if all three are odd, change entry C.

And, if all three are even, make no changes.

How useful is this process? Suppose information is being transmitted electronically, and there is one chance in 10 that an error will be made on any given symbol. Then the probability of a four-symbol binary message being received correctly is $0.9^4$, or about 0.656. So roughly one in every three messages will be received incorrectly. If there are seven symbols, the chance of no errors is $0.9^7$, approximately 0.478. The chance that the first symbol will be received incorrectly but the rest will be correct is $0.9^6 \times 0.1$, and it is the same for each other symbol, so the chance of exactly one error is $0.9^6 \times 0.1 \times 7$, or 0.372. So the probability of at most one error is 0.85. In those cases the Venn diagram method returns the correct message, so Venn diagram encoding improves the chance of receiving the message correctly from about 2 in 3 to about 5 in 6.

However, most electrical transmissions are far more reliable than that. If the chance of an error in any symbol is 1 in 100, the probability of an error in a four-digit transmission is about 1 in 25, while the probability that the message would be decoded correctly using Venn diagram decoding is 0.998—the chance of a mistake in a message is 1 in 500.

Of course, errors are more likely in some transmissions. An example is transmissions from probes sent to the moon and planets. For these cases, methods are available that correct far more than single errors.

## 10.3   Binary Codes

We now discuss (binary) codes in general.

Any binary string (that is, string of 0s and 1s) can be used to encode information, so we will call any such string a *binary codeword*. The number of symbols in the string is called its *length*. The *distance* between two binary codewords is defined to be the number of places in which they differ. For example, 1010111 and 1100101 are binary codewords of length 7, and the distance between them is 3, because they differ in the second, third and sixth places.

By a *binary code* we mean a set of binary codewords. Usually we shall just say "code"; while codes based on a set with more than two symbols are used, binary codes are by far the most common. In most cases, we shall require all words in a code to be of same length (but there are a few important exceptions).

The minimum distance between any two codewords in a code (that is, the smallest of all differences between two codewords) is called the *minimum distance* of the code.

As an example, here is a code with four words:

$$A = 1010101$$
$$B = 0101010$$
$$C = 1110000$$
$$D = 0001111$$

The distances between the words are

| | |
|---|---|
| A to B: 7 | B to C: 4 |
| A to C: 3 | B to D: 3 |
| A to D: 4 | C to D: 7 |

So the minimum distance is 3.

The *weight* of a codeword is the number of 1s it contains. That is, it is the distance from $000\ldots00$. The weight of a *code* is the minimum weight of all its non-zero words. If $000\ldots$ is in the code, you do not include it in calculating the minimum weight.

In the preceding example:

$$A = 1\ 0\ 1\ 0\ 1\ 0\ 1 \quad ; \text{weight is} : 4$$
$$B = 0\ 1\ 0\ 1\ 0\ 1\ 0 \quad ; \text{weight is} : 3$$
$$C = 1\ 1\ 1\ 0\ 0\ 0\ 0 \quad ; \text{weight is} : 3$$
$$D = 0\ 0\ 0\ 1\ 1\ 1\ 1 \quad ; \text{weight is} : 4.$$

This code has weight 3.

The *binary sum* or *join* or *bit sum* or simply *sum* of two binary sequences is defined as follows.

To add symbols, use the "cancellation" rule

$$0+0 = 0 \quad 0+1 = 1$$
$$1+0 = 1 \quad 1+1 = 0$$

In other words, the sum is 1 if the symbols are different and 0 if they are the same. To add *sequences* (for example, codewords) add terms separately, but there is no carry:

$$
\begin{array}{r}
1\ 0\ 1\ 0\ 1\ 0\ 1\ 1 \\
+\ 1\ 1\ 1\ 0\ 0\ 0\ 1\ 0 \\
\hline
0\ 1\ 0\ 0\ 1\ 0\ 0\ 1
\end{array}
$$

So the distance between two sequences is the weight of their sum. The sum of any binary string with itself is the string $000\ldots$, and as one would expect the distance between a string and itself is zero.

Some writers use different symbols such as $\oplus$ or $\odot$ to represent binary sums, but no confusion should arise if we simply use the ordinary plus sign. The binary sum obeys the ordinary associative and commutative laws. We shall denote the string of all zeros as $O$; if $A$ is any binary string, $A + O = A$ and $A + A = O$.

Suppose $A$ and $B$ are any two binary sequences of the same length and $C$ is their sum: $A + B = C$. Then $A + B + C = C + C = O$. Moreover, $A + C = A + O + C = A + B + B + C = O + B = B$ and $B + C = A$.

A very important example is the *Hamming code* of length 7. It is derived from the 16 possible strings of length 4 (of 0s and 1s). To form a 16-word binary code with words of length 7, do Venn diagram encoding on each string.

| String | Codeword | String | Codeword |
|--------|----------|--------|----------|
| 0000 | 0000000 | 0001 | 0001011 |
| 0010 | 0010111 | 0011 | 0011100 |
| 0100 | 0100101 | 0101 | 0101110 |
| 0110 | 0110010 | 0111 | 0111001 |
| 1000 | 1000110 | 1001 | 1001101 |
| 1010 | 1010001 | 1011 | 1011010 |
| 1100 | 1100011 | 1101 | 1101000 |
| 1110 | 1110100 | 1111 | 1111111 |

This code—which is called the *Hamming code* of length 7—is an example of a linear code. The 16 codewords are precisely the strings that we would assume were the intended codewords when we decode a string using Venn diagram decoding.

A *linear code* is one in which the binary sum of any two of the codewords is also a codeword. Clearly all the words in a liner code must be of the same length (or else the sum will not always be defined).

Many linear codes are derived by:

- taking all strings of some fixed length;
- adding check digits;
- possibly changing the order of the digits in some consistent way.

However, not all linear codes are derived in this way.

Here are two important properties of linear codes. First, the all-zero word $000\ldots$ is always one of the codewords. Second, the minimum distance of the code is equal to the weight of code.

*Nearest-neighbor decoding* is the following process: find the codeword whose distance from the received message is smallest; decode to that codeword.

Venn diagram decoding is an example of nearest-neighbor decoding. When we explained how to derive the intended message from any seven-digit binary string, we could always derive a codeword by leaving the string unchanged or making one change.

If there were two errors, Venn diagram decoding would give the wrong answer. But it would at least see that there had been an error. If three errors were made, you might not see the error (because it is possible to get another codeword by this process). That is, Venn diagram decoding *(detects and) corrects 1 error in a word*, and *detects any 2 errors*. We say it is a *single error correcting, double error detecting* code. In a single error correcting code, if the string is not a codeword, then the position whose symbol must be changed in order to make it a codeword is unique.

If a code has minimum distance $t$, it will detect any $(t-1)$ errors (or fewer) and correct any $(t-1)/2$ errors (or fewer). Alternatively, we could say that a code with minimum distance $t$ will detect up to $(t-1)$ errors and correct up to $(t-1)/2$ errors.

**Sample Problem 10.2**  *Consider the code with four words:*

$$
\begin{aligned}
A &= 1\ 0\ 1\ 0\ 1\ 0\ 1 \\
B &= 0\ 1\ 0\ 1\ 0\ 1\ 0 \\
C &= 1\ 1\ 1\ 0\ 0\ 0\ 0 \\
D &= 0\ 0\ 0\ 1\ 1\ 1\ 1
\end{aligned}
$$

*How many errors can this code correct? How many can it detect?*
**Solution.** The code has minimum distance 3. So it detects any 2 errors, corrects any 1 error.

There is no such thing as half an error. So, if your arithmetic tells you that a code can detect up to $2\frac{1}{2}$ errors, the maximum number it will actually detect is 2.

**Sample Problem 10.3**  *Consider the code with words:*

$$
\begin{aligned}
A &= 1\ 1\ 1\ 1\ 0\ 0\ 0\ 0 \\
B &= 1\ 1\ 0\ 0\ 1\ 1\ 0\ 0 \\
C &= 0\ 0\ 1\ 1\ 0\ 0\ 1\ 1 \\
D &= 0\ 0\ 0\ 0\ 0\ 0\ 0\ 0
\end{aligned}
$$

*What is the most errors can this code correct? What is the most it can detect?*
**Solution.** The code has minimum distance 4. So it detects any 3 or fewer errors. It corrects "up to 1.5 errors", but as you can only have whole numbers of errors, it can only detect 1 error.

The Hamming code has one particular property that is not true of many codes. Given any string of seven binary digits, there is exactly one codeword that is distance 0 or 1 from the string: either the string is a codeword, or there is a unique codeword into which it can be converted by changing one symbol. We shall call this the *unique correction property*.

## 10.4    Variable-Length Codes

Linear codes are necessarily fixed-length. Many codes are like this, but some codes use different length words.

One well-known example is the Morse code. It was designed for sending English-language messages by telegraph, and it was desirable to make the transmissions as short as possible. For this reason, those letters that occur most commonly in English have shorter representations.

The Morse code is binary. The two symbols are the dot and the dash, where the dot is a short transmission and the dash is a longer one, as long as three dots. Each dot or dash is followed by a short silence, equal to the dot duration; the letters of a word are separated by a silence equal to three dots (one dash), and two words are separated by a space equal to seven dots. For convenience, we shall represent the dot by 0 and the dash by 1. Table 10.1 shows the Morse code for letters, and also the approximate frequency of letters in English. Table 10.2 shows how numbers are represented in the code.

**Table 10.1**  Representations of letters in Morse code, and letter frequencies in English

|   | Code | % |   | Code | % |   | Code | % |   | Code | % |
|---|------|-----|---|------|-----|---|------|-----|---|------|-----|
| A | 01   | 8   | B | 1000 | 1.5 | C | 1010 | 3   | D | 100  | 4   |
| E | 0    | 13  | F | 0010 | 2   | G | 110  | 1.5 | H | 0000 | 6   |
| I | 00   | 6.5 | J | 0111 | 0.5 | K | 101  | 0.5 | L | 0100 | 3.5 |
| M | 11   | 3   | N | 10   | 7   | O | 111  | 8   | P | 0110 | 2   |
| Q | 1101 | 0.25| R | 010  | 6.5 | S | 000  | 6   | T | 1    | 9   |
| U | 001  | 3   | V | 0001 | 1   | W | 011  | 1.5 | X | 1001 | 0.5 |
|   |      |     | Y | 1011 | 2   | Z | 1100 | 0.25|   |      |     |

**Sample Problem 10.4** *How is the phrase "bring food" represented in Morse code?*

**Solution.** 1000 010 00 10 110        0010 111 111 100

**Your Turn.** How is "delayed" represented in Morse code?

The Morse code is an example of a *variable-length* code. Some variable-length codes use a break to symbolize the end of a word, either a physical break in transmission (like the Morse code) or an "end-of-line" symbol; in others, the codewords can include information that signals the end of the word. For example, if every codeword ends in 00, and there are never two consecutive 0s in a word, each 00 would tell you when a word ends.

Another important example is the genetic code. Genes are built from four nucleotides: adenine, thymine, Guanine, and cytosine. A gene can be specified by a sequence of these, usually represented by the initials A, T, G, and C.

Suppose we want to use a binary code to represent a genetic string. One possibility is to use the codewords 00 for A, 01 for C, 10 for G, and 11 for T.

**Table 10.2**  Representations of numbers in Morse code

| 1 | 01111 | 2 | 00111 | 3 | 00011 | 4 00001 | 5 00000 |
|---|-------|---|-------|---|-------|---------|---------|
| 6 | 10000 | 7 | 11000 | 8 | 11100 | 9 11110 | 0 11111 |

For example, AACAGTAAAC would be represented by 00000100101100000001 (20 symbols).

It happens that A is most common in genetic strings, then C, then T and G. We could try the following encoding:

0 for A,    10 for C,    110 for T,    111 for G.

Then AACAGTAAAC is represented by 0010011111000010 (16 symbols). We shall call this the *genetic code*. Notice that, in the suggested encoding method, every time 111 occurs, it is the codeword for G, and every time a 0 occurs, it is at the end of a codeword. Moreover, the only way a codeword can end is with 0 or 111. So here is a technique for decoding the genetic code. First, put commas after each occurrence of 0 or 111. Then decode the strings between the commas. For example, the string

00110010111011011000010

is written

0,0,110,0,10,111,0,110,110,0,0,0,0,10

which decodes to

AATACGATTAAAAC.

## 10.5   The Hat Game

We conclude this chapter with a puzzle whose answer may surprise you.

We consider a game show with a team of three contestants; call them players 1, 2, 3 in some order. (Alphabetically, maybe.) The contestants can meet before the show in order to discuss strategies. But then they are taken to separate dressingrooms, and each is given a hat. They cannot see the hats that they are allocated; the player only knows that the hat is either red or black. For any given player, there is an equal chance that red or black will be allocated—for example, a coin is tossed for each player, out of the player's sight; heads means red, tails means black.

The players then go onto a stage; they are too far apart to communicate, but each player can see the colors of the teammates' hats. Then the players vote simultaneously. (A "vote" consists of the statement "my hat is black" or the statement "my hat is red.") There is no way for any of them even to give a hint to their partners about the hat colors.

The players do not have to vote; at their turn they can pass instead. But the requirement is that at least one player must vote correctly, and no player can vote incorrectly. so, in order to win, every player must either pass or get his or her own hat color correct.

Obviously, the players will talk before the game, and try to come up with a strategy. After all, if all three players vote, and each would be correct about half the time, there would be only one chance in eight of winning the prize. One obvious plan is to appoint a spokesman, who will say "my hat is red," while the others pass. This has a 50% chance of being correct. But can they do better?

Yes, they can! The players work on the assumption that not all the hats are the same color. Each player looks at her two partners. If their hat colors are both red, hers is black; if both black, hers is red; if they are different, she must pass. For example, if the colors are red, black, red in the order of the players, then both 1 and 3 see two different colors, so they pass. Two sees both red, so votes "black." The players win.

Consider the eight possibilities.

- BBB All three vote R. They lose.
- RBB 1 votes R; others pass. They win.
- BRB 2 votes R; others pass. They win.
- BBR 3 votes R; others pass. They win.
- BRR 1 votes B; others pass. They win.
- RBR 2 votes B; others pass. They win.
- RRB 3 votes B; others pass. They win.
- RRR All three vote B. They lose.

So they win in six of the eight cases, or 75%.

To understand what we have done here, suppose we represent red hats by 1 and black hats by 0. The actual hat configuration is represented by a binary string of length 3. Think of them as possible codewords in a particularly simple binary code with only two messages, 1 and 0. To encode a message, simply send the message three times. If you receive 110 in this code, you know that there has been either one error (the intended transmission was 111) or two (it should have been 000). In nearest-neighbor decoding, you would assume there was only one error, so it was intended to send 111 and the message was 1.

Now suppose you are told that a codeword was sent incorrectly, but you can only see two of the characters. If those two characters are different, you cannot tell what was the third character sent, but if two were the same then the third must have been different from them.

We turn this coding problem into a game for three players as follows. A three-symbol binary string is transmitted at random; the strings are equally likely. Each player sees two of the symbols (a different pair for each player), but cannot see the third. They are asked if they know the missing symbol. They can either identify it or pass; the players win if at least one identifies the symbol correctly and none identify it incorrectly.

This code game corresponds to the hat game. A player receiving a red hat in the hat game is the same as if that player cannot see a 1 in the code game. The players work on the assumption that the string was not a codeword. If the string transmitted was actually a codeword, all three players would think they knew the missing symbol but would be wrong. If there was a transmission error, one player would give a correct answer while the other two would pass. So the players would lose if the string was a codeword and win if it was not. Since only two of the eight possible strings are codewords, they would win in 75% of cases.

We can generalize the hat game to seven players by using the Hamming code. Again, a red hat corresponds to a 1 in the code and a black hat to an R. The players are ordered, and each player can see the other hats and construct a seven-digit string with one digit missing—the one corresponding to the player's own hat. She then asks, "what would the missing symbol be if this is *not* a codeword?" It can be shown that, if the string is not a codeword, exactly one player can answer this question; the others will pass, and the team will win. If the original string *was* a codeword, all the player will make a wrong answer. Since there are 16 possible codewords and $2^7 = 128$ possible strings, the probability of winning this game is 7/8.

# Multiple Choice Questions 10

1. Use the Venn diagram method to determine the code word of the message 1001.
   A.   1001111                           B.   1001101
   C.   1001001                           D.   1001010

2. Use the Venn diagram method to determine the code word of the message 1010.
   A.   1010111       B.   1010011       C.   1010001       D.   1010110

3. Use the Venn diagram method to decode the received word 1111001.
   A.   1011            B.   0111            C.   0010            D.   1111

4. Use the Venn diagram method to decode the received word 1011001.
   A.   1011            B.   1001            C.   1010            D.   0100

5. Add the binary sequences 1001010 and 1010010. How many digits 1 are in the sum?
   A.   2          B.   3          C.   4          D.   Some other answer.

6. What is the distance between the words 1111000 and 1010101?
   A.   1                    B.   2                    C.   3                    D.   4

7. A code consists of the words $\{1010, 0101, 1111, 0000, 1001\}$. What is the weight of this code?
   A.   0                    B.   1                    C.   2                    D.   4

8. Consider the code $\{101010, 010101, 111111, 000000\}$. What is the weight of this code?
   A.   0                    B.   1                    C.   3                    D.   6

9. For the code consisting of the two words 00000 and 11111, how many errors would have to occur during transmission for a received word to be decoded incorrectly, using nearest-neighbor decoding?
   A.   5          B.   3 or more          C.   4 or more          D.   2 or more

10. If a code has minimum distance 4, what is the maximum number of errors it can correct?
    A.   1                    B.   2                    C.   3                    D.   4

11. What is the distance between received words 1001010 and 1010010?
    A.   1                    B.   2                    C.   3                    D.   4

12. Consider the code $\{1100, 1010, 1001, 0110, 0101, 0011\}$. Which of the following is a true statement?
    A.   The code can detect some but not all single-digit errors
    B.   The code can detect and correct any single-digit error
    C.   The code can detect any single-digit error and correct some but not all single-digit errors
    D.   The code can detect any single-digit error, but cannot correct any single-digit error

## Exercises 10

1. Determine the codewords that would be transmitted for the messages 1000 and 0100, using the Venn diagram method.
2. Determine the codewords that would be transmitted for the messages 1110 and 0101, using the Venn diagram method.
3. Determine the codewords that would be transmitted for the messages 1011 and 1001, using the Venn diagram method.
4. Encode the message 1111 1101 1011 0011.
5. Use the Venn diagram method to decode the received word 1001010.
6. Use the Venn diagram method to decode the received message 1001011 1000110 1010000.
7. Suppose the Venn diagram message 1110 is received as 1110001. Will the original message be recovered? Why, or why not?
8. What is the binary sum of the sequences 10010101 and 10100011?
9. What is the distance between words 11100011 and 10110110?
10. What are the weights of the two words given in the preceding question?
11. Consider the code {000000, 111111}.
    (i)  What is the weight of the code?
    (ii)  How many single-digit errors will the code detect?
    (iii)  How many single-digit errors will the code (detect and) correct?
12. Suppose a code has minimum distance 2. Show that the code cannot correct errors (using nearest-neighbor).
13. Consider the code with eight codewords

$$A \; = \; 1100, \;\; B \; = \; 1010, \;\; C \; = \; 1001, \;\; D \; = \; 0110,$$
$$E \; = \; 0101, \;\; F \; = \; 0011, \;\; G \; = \; 1111, \;\; H \; = \; 0000.$$

    (i)  Prove that this is a linear code.
    (ii)  What is the weight of this code?
    (iii)  What is the minimum distance of this code?
    (iv)  How many errors can this code detect and correct, using nearest-neighbor decoding?
14. Repeat the preceding question for the code with words

$$A \; = \; 11001100, \;\; B \; = \; 10101010, \;\; C \; = \; 10011001,$$
$$D \; = \; 01100110, \;\; E \; = \; 01010101, \;\; F \; = \; 00110011,$$
$$G \; = \; 11111111, \;\; H \; = \; 00000000.$$

15. A code is constructed as follows. The messages are the 16 binary strings of length 4. If the message is $xyzt$, then the codeword transmitted is $xyztabcd$, where $a = y+z+t, b = x+z+t, c = x+y+t$ and $d = x+y+z$ (binary sums).
    (i)  Using this code, what are the codewords corresponding to the messages 1111 and 0000?
    (ii)  What are the codewords corresponding to the messages 1001, 1110 and 0001?

(iii) What are the weights of the words in this code? What is the weight of the code?

(iv) How many errors can this code correct?

(v) Decode the messages 11110111 and 00010011 using nearest-neighbor decoding.

16. What is the Morse code encoding for "Come here tomorrow"?

17. You receive the following in Morse code. What was the message?

$$1000010001101001010001$$

18. Encode AATCACTGAACA in the genetic code.

19. Decode the genetic code string 00110100101101111100100.

20. Decode the genetic code string 011010111011011010111010011010111101000.

# Chapter 11
# Cryptography

## 11.1  Secret Writing

The idea of *encoding* data arises in many ways. We have already seen the way in which information is encoded in identification numbers, such as the ISBN or driver's license numbers. Another important form of encoding that we saw is the addition of check digits for error correction, which is used both in identification numbers and in sending data.

In this chapter we shall look at another reason for encoding—secrecy. Sometimes you encode something and only tell certain selected people how to decode it. This form of secret encoding is called *encryption*, and the study of it is *cryptography*. It has been widely used in military operations, but nowadays it is very common in commerce. For example, the use of a PIN number with your bank card is a form of encryption.

The original message, in everyday language, is often called a *plaintext*, and the encoded version is the corresponding *ciphertext*. ("Cipher" is another word for "code".) The process of moving from ciphertext to plaintext (the opposite of encryption) is called *decryption*. The special information required by your ally to decrypt your message is called a *key*.

Suppose you send a secret message. If your enemy intercepts it, he will try to figure out what it means by working out how you did the encoding—usually, he tries to find the key. This is called *cryptanalysis*, or codebreaking. The study of cryptography and cryptanalysis is collectively called *cryptology*.

There are many mystery stories in which cryptography takes an important part. If you want to read some of these, I recommend *The Adventure of the Dancing Men*, a Sherlock Holmes story by Sir Arthur Conan Doyle, Edgar Allen Poe's *The Gold Bug*, or *The Nine Tailors* by Dorothy L. Sayers.

W.D. Wallis, *Mathematics in the Real World*, DOI 10.1007/978-1-4614-8529-2_11,
© Springer Science+Business Media New York 2013

## 11.2   Physical Secrecy

Some 2500 years ago the Spartans during their wars used a special device to transport their secret messages. They took a staff called a *scytalee* (pronounced "sitterlee"), and wound a long, thin strip of parchment around it at an angle. They then wrote their message on the parchment, one letter on each width of parchment. When the strip is unwound, the letters appear to be completely random (try it yourself). But when the roll is wound again around another staff of equal diameter to the original staff, the message makes sense, but nonsense appears if you use the wrong size staff. The word scytalee is also used for the roll after it is wrapped.

As a very simple example, suppose you wanted to send the message "help is Coming." You would wrap a piece of parchment around your staff, and write your message on it in a few lines. Suppose you decided to write it in three lines: the first line would be "HELP", the second "ISCO" and the third "MING" (there is no need to include spaces—in fact, if you include spaces between words, it might help your enemy to decode your message). We'll call the number of lines, 3 in this case, the *circumference* of the scytalee. You end up with something like Fig. 11.1. The strip will actually say "HIMESILCNPOG," but when your allies wrap it around their own staff (the same diameter as yours, so that the scytalee comes out to have circumference 3) they get the welcome news "HELPISCOMING."

However, suppose your enemy intercepts the message. If she uses a staff of the same diameter, you are in trouble; but if she uses a slightly larger staff, so that the circumference is 4, she will see the picture in Fig. 11.2, and try to interpret "HSNIIPMLOECG".

Nowadays this would not be a very secret method of writing, as it can be easily deciphered by moving around the edges of the roll; but very few people knew how to read at the time, and this ensured the secrecy of the scytalee method. Even if your enemy could read, if she did not know about the scytalee method, she would probably think the strip was nonsense, or else think it was in some strange language.

**Sample Problem 11.1**   *Suppose you receive the scytalee message*

*SYBLCRESEERACHTAYPUOHIPHRUEMTYILSOOTDOFG*

*How would you decode it?*

**Solution.** To decode this we try various different circumferences. If we try circumference 5 (and rearrange our message in five rows) we get complete nonsense, as follows.

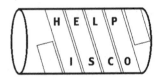

**Fig. 11.1**   A scytalee

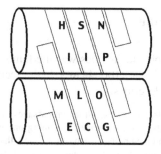

**Fig. 11.2** Misinterpreting the scytalee

$$
\begin{array}{ccccccccc}
S & R & R & A & H & U & I & ! & G \\
Y & E & A & Y & I & E & L & T \\
B & S & C & P & P & M & S & D \\
L & E & H & U & H & T & O & O \\
C & E & T & O & R & Y & O & F \\
\end{array}
$$

However, if we arrange it in six rows, the message becomes clear.

$$
\begin{array}{ccccccc}
S & E & C & U & R & I & T \\
Y & S & H & O & U & L & D \\
B & E & T & H & E & S & O \\
L & E & A & I & M & O & F \\
C & R & Y & P & T & O & G \\
R & A & P & H & Y & ! \\
\end{array}
$$

**Your Turn.** Suppose you receive the scytalee message

$$CYRDIEAAOETSSGNUAHMSE$$

How would you decode it?

Other physical cryptography techniques were used. Histiacus, a Greek ambassador in Persia, sent secret information home to his country by shaving a slave's head, branding a message on his skull and letting the hair grow back. He then sent him to Greece. Others used to drug their slaves and brand messages on their backs. The slaves knew nothing about the important messages they were carrying. In both cases, it was unlikely that the slave would live very long after the message was delivered.

## 11.3   The Caesar Cipher

An early user of cryptographic techniques was the famous Roman commander-in-chief and statesman Julius Caesar (100–44 BC). In the second century AD Suetonius (*The Twelve Caesars* I,56) describes:

> his private letters to friends, the more confidential passages of which he wrote in cipher: to understand their apparently incomprehensible meaning one must number the letters of the alphabet from 1 to 22, and then replace each of the letters that Caesar has used with the one which occurs three numbers lower—for instance, *D* stands for *A*.

The numbers 1 to 22 are used above because the Romans used only 22 letters in their alphabet. But, for convenience, let's pretend Caesar used the English alphabet and used the same alphabetical order as we do. We obtain Caesar's cipher if we write the ciphertext alphabet beneath the cleartext alphabet, shifted three positions to the left. Notice that the alphabet wraps around at the end.

$$\text{Cleartext:} \quad a \quad b \quad c \quad ... \quad w \quad x \quad y \quad z$$
$$\text{Ciphertext:} \quad D \quad E \quad F \quad ... \quad Z \quad A \quad B \quad C$$

We encipher a cleartext letter by replacing it by the letter beneath it. For instance, the message *"the cleartext"* becomes *WKH FOHDUWHAW*. (Caesar didn't use the trick of omitting blanks.) Deciphering is equally easy; we replace a ciphertext letter by the cleartext letter above it. So *OHWWHU* is deciphered to the message *"letter."*

Caesar chose a three-letter skip for no particular reason; one could of course get the ciphertext alphabet by shifting any number of positions. The key is the number of places you shift; Caesar's key was 3. Since the English alphabet consists of 26 letters, we obtain exactly 26 such ciphers; they are called additive ciphers. The one with key 0 (or 26) is no good—the ciphertext is exactly the same as the plaintext—so there are 25 useful additive ciphers in English.

In the movie *2001: A Space Odyssey*, the computer is called HAL; this can be obtained by moving the letters IBM one step to the left (or 25 steps to the right). The producers claimed that this was a coincidence.

## 11.4   The Vigenère Method

Vigenère invented an encryption method that combines several additive ciphers. In modern terminology we could say that the key is a string of numbers, each one between 0 and 25. The easiest way to remember this is to convert it into a word, using $A = 0, B = 1, \ldots$. For example, the key $1, 17, 14, 22, 13$ is equivalent to the keyword BROWN. Then the plaintext is written out, with the keyword underneath it, repeated as necessary. For example, to encrypt the message *the computer has become very useful* we write

```
THECOMPUTERHASBECOMEVERYUSEFUL
BROWNBROWNBROWNBROWNBROWNBROWN
```

Now each letter is encrypted using the additive cipher whose key is below it. In the example, *T* would become *U*, *H* would become *Y*, and so on. The message becomes

```
UYS YBNGIPRS YOO OFTCIR WVFU HTVTQY.
```

In fact, the ciphertext would not usually be broken into words. You would probably receive

```
UYSYBNGIPRSYOOOFTCIRWVFUHTVTQY,
```

or perhaps it would arrive arbitrarily broken into four- or five-letter groups in order to make copying errors less likely:

```
UYSY BNGI PRSY OOOF TCIR WVFU HTVT QYZZ
```

where the last two letters have been added simply to make the last set of letters the right size. (Presumably the recipient of a message will have no trouble ignoring a couple of nonsense letters at the end.)

## 11.5 Substitution Ciphers

In an additive cipher, each letter is changed into another fixed letter. For example, in the Caesar cipher, every *A* becomes a *D*, every *B* becomes an *E*, and so on. To generalize this, we could make a fixed substitution for each letter, but use a more complicated rule than simple addition. You could write out the complete alphabet, and the list of substitutions, like

| Cleartext:  | *a* | *b* | *c* | *d* | *e* | *f* | *g* | *h* | *i* | *j* | *k* | *l* | *m* |
|-------------|-----|-----|-----|-----|-----|-----|-----|-----|-----|-----|-----|-----|-----|
| Ciphertext: | *D* | *A* | *F* | *B* | *Q* | *K* | *M* | *C* | *Z* | *Y* | *E* | *G* | *H* |
| Cleartext:  | *n* | *o* | *p* | *q* | *r* | *s* | *t* | *u* | *v* | *w* | *x* | *y* | *z* |
| Ciphertext: | *W* | *X* | *P* | *R* | *S* | *I* | *T* | *L* | *J* | *N* | *V* | *U* | *O* |

In this code the message *"an elephant"* becomes *DWQGQPCDWT*. (It is usual to omit spaces between words.) There are arithmetical ways to break codes such as this, based on the frequency of letters. For example, *E* is the most common letter in English, so if you think your message is written in English, try putting *E* for the most commonly occurring letter. This method only works for reasonably long messages; for short ones, you need the key (the whole substitution table).

One simplification of this method involves the use of a keyword. The second row of the table is constructed by writing first the keyword and then the rest of the alphabet in order. The keyword should not be too short; it should contain no repeated letters, and preferably one of the letters should be rather late in the alphabet. For example, keyword *CRAZY* produces the table

| Cleartext:  | *a* | *b* | *c* | *d* | *e* | *f* | *g* | *h* | *i* | *j* | *k* | *l* | *m* |
|-------------|-----|-----|-----|-----|-----|-----|-----|-----|-----|-----|-----|-----|-----|
| Ciphertext: | *C* | *R* | *A* | *Z* | *Y* | *B* | *D* | *E* | *F* | *G* | *H* | *I* | *J* |
| Cleartext:  | *n* | *o* | *p* | *q* | *r* | *s* | *t* | *u* | *v* | *w* | *x* | *y* | *z* |
| Ciphertext: | *K* | *L* | *M* | *N* | *O* | *P* | *Q* | *S* | *T* | *U* | *V* | *W* | *X* |

**Sample Problem 11.2**  *Use the keyword CRAZY to encrypt the message "send troops tomorrow."*
**Solution.** Using the table, *s* becomes *P*, *e* becomes *Y*, and so on. The ciphertext is *PYKZQOLLMPQLJLOOLU*.
**Your Turn.** Use the keyword *CRAZY* to encrypt the message "attack now."

## 11.6  Modular Arithmetic

Suppose you look at a clock and want to know, "what time will it be in five hours?" If it is currently 10 o'clock, you don't say "15 o'clock," you say "3 o'clock." You might say "AM" or "PM," but that is extra information, not shown on a standard clockface. In the same way, 44 h from now it will be 6 o'clock; what day, and whether it will be morning or evening, is not given by clockface arithmetic.

What you do when interpreting a clock is to subtract multiples of 12 when necessary. In fact, all you are concerned with is the *remainder* after division by 12. For example, $10 + 44 = 54 = 4 \times 12 + 6$; we ignore the 12s and say "6 o'clock."

Another way of expressing clockface arithmetic is to say calculations are carried out *modulo 12*, and we might write

$$44 + 10 = 6 (\text{mod } 12).$$

In order to distinguish between modular arithmetic and ordinary arithmetic, it is usual to write this as $44 + 10 \equiv 6 (\text{mod } 12)$, although most people still say "equals." The number 12 is called the *base* or the *modulus*. We usually express the answer to a modulo 12 problem as one of the numbers in the standard range $0, 1, 2, \ldots, 11$, although we could equally well write 12 instead of 0. (In the particular case of telling the time, we say "12 o'clock," not "0 o'clock.")

Another method of telling time is the military, or 24-h clock. This indicates the difference between morning and evening, and is in fact a modulo 24 arithmetic. (In the military clock we say "00:30 h," not "24:30 h.")

In the preceding chapter we used binary arithmetic, the arithmetic of even and odd. It is about remainders modulo 2. In that case the addition table is

$$0+0=0 \qquad \text{Even} + \text{Even} = \text{Even}$$
$$0+1=1 \qquad \text{Even} + \text{Odd} = \text{Odd}$$
$$1+0=0 \qquad \text{Odd} + \text{Even} = \text{Odd}$$
$$1+1=0 \qquad \text{Odd} + \text{Odd} = \text{Even}$$

Binary arithmetic comes up in many electrical applications. As an example, think about turning a light on or off. On many lamps, there is a knob that you twist clockwise. If the light is on, twisting the knob changes it to "off." Similarly twisting from "off" gives "on." If 1 means "on" and 0 means "off," and if the process of turning the knob is interpreted as "add 1" (which makes sense if you think of the lamp being off originally), you have

$$0+1=1, \quad 1+1=0,$$

which is precisely arithmetic modulo 2. The Venn diagram encoding method could be expressed as follows: "the entries E, F, and G are chosen so that the sum of entries in each circle is 0 (modulo 2)."

Additive ciphers use arithmetic modulo 26 (provided the language uses the English alphabet). The Caesar cipher could be expressed as "add 3 modulo 26," and decoding is "subtract 3 modulo 26," where the letters are interpreted as numbers ($A = 1$, $B = 2$, and so on up to $Z = 26$).

This sort of modular arithmetic can be carried out with any integer as modulus. We can do addition, multiplication, raising to powers, or any arithmetical operation.

For example, in arithmetic modulo 5, the calculation "$3^3 + 2$" is carried out as follows. $3^3 = 27$ and $27 + 2 = 29$. Now $29 \equiv 4 \pmod 5$ (this process of getting rid of 5s and coming down to a number in the standard range is called *reducing modulo the base*). So the answer is 4. Notice that you can reduce at any stage: the above calculation could have gone $3^3 = 27 \equiv 2 \pmod 5, 2 + 2 = 4$.

**Sample Problem 11.3** *Solve the equation* $x^2 \equiv 1 \pmod 5$.
**Solution.** The easiest technique in this case is to test all the possibilities in the standard range: there are only five of them. We find $0^2 = 0$, $1^2 = 1$, $2^2 = 4$, $3^2 = 9 \equiv 4$, and $4^2 = 16 \equiv 1$. So the answer is "$x = 1$ or 4."

One well-known application of modular arithmetic is deciding whether a number is divisible by 9. When we write numbers in the usual way, we observe that $10 \equiv 1 \pmod 9$, $100 \equiv 1 \pmod 9$, $1000 \equiv 1 \pmod 9$, and so on. Moreover, $200 \equiv 2 \pmod 9$, $300 \equiv 3 \pmod 9$, and so on. We can use this to calculate the remainder on division by 9: for example,

$$3,182 = 3,000 + 100 + 80 + 2 \equiv 3 + 1 + 8 + 2 \equiv 14 \equiv 1 + 4 = 5 \pmod 9.$$

So 3,182 leaves remainder 5 on division by 9. We simply add the digits.

## 11.7   The RSA Scheme

Modular arithmetic is used in a cryptographic system invented by Ronald L. Rivest, Avi Shamir and Leonard M. Adelman, which is usually called RSA.

We shall describe their system by giving a simple example. Remember that a *prime* is a number whose only divisors are itself and 1. Since the divisor 1 can never be avoided, we shall call divisors bigger than 1 *proper*, so a prime is a number whose only proper divisor is itself. Your friend, the person to whom you wish to send a secret message, must choose two primes—let's call them $p$ and $q$—and another number, $r$, with the following property: no proper divisor of $r$ can also be a divisor of $p-1$ or $q-1$. The *least common multiple* of $p-1$ and $q-1$—the smallest number such that both $p-1$ and $q-1$ divides it—will also be important; we shall denote it $m$. Your friend also has to find a number $s$ such that $r \times s \equiv 1 \pmod{m}$.

For the purposes of our example, suppose your friend chose $p = 7$ and $q = 17$, so that $m = 48$. Then $r$ can have no proper divisor that also divides 6 or 16. 2, 3, 4 and 6 are impossible, but 5 and 7 are allowable. Let us assume she chose 5. The number $s$ must satisfy $5 \times s \equiv 1 \pmod{48}$. So $5 \times s$ must be one of the numbers $1, 49, 97, 145, 193, \ldots$. Since the product must be divisible by 5, your friend selects 145, and $s = 29$.

Say you wish to send the message "CAP." You first convert the message to numbers by replacing A by 1, B by 2, ..., and Z by 26. The plaintext CAT becomes 3 1 16. To convert a number from plaintext to ciphertext, you raise it to the power $r$ and then reduce the answer modulo $pq$. Remember, $r = 5$, $pq = 7 \times 17 = 119$.

$3^5 = 243 \equiv 5 \pmod{119}$, so 3 is encoded 5;

$1^5 = 1 \equiv 1 \pmod{119}$, so 1 is encoded 1;

$16^5 \equiv 67 \pmod{119}$, so 16 is encoded 67 (hall point out below how this was calculated).

Therefore the ciphertext for "CAP" is 5 1 67.

When your friend receives the cryptic message 5 1 67, she raises each number to the power $s$, which is 29 in this case. $5^{29}$ is a very large number, bigger than 180 million million million, but it is easily shown to be 3 modulo 119 (see the Sample Problem below; of course, computer programs are available that do these calculations automatically). So 5 is decoded as 3. Similarly 1 decodes to 1 and 67 to 16, and "CAP" is recovered.

We chose "CAP" to show you that even in a very small example, big numbers arise. One could find $16^5 = 1,048,576$ on a calculator directly, but other problems could involve larger numbers, possibly too big for your calculator to display with all the digits. An alternative method is to reduce modulo the base in the middle of the calculation. Observe that

$$16^5 = 16^2 \times 16^2 \times 16,$$

Now $16^2 = 256$, and by subtracting 119 (twice) we find $256 \equiv 18 \pmod{119}$. Then

$$18 \times 256 \equiv 18 \times 18 \equiv 324 \equiv 86 (\text{mod } 119).$$

Or you could start from $18 \times 256 = 4608$. To reduce mod 119, first divide and find $4608/119 = 38.72\ldots$. Subtract 119 from 4608 thirty-eight times (actually, perform $4608 - (38 \times 119)$ on the calculator) and get 86. Then

$$86 \times 16 = 1376 \equiv 67 (\text{mod } 119).$$

**Sample Problem 11.4**   *Calculate $5^{29}$ modulo* 119.
**Solution.** First observe that $5^2 = 25$, $5^4 = 625 \equiv 30$, $5^8 \equiv 30^2 \equiv 67$, $5^{16} \equiv 67^2 \equiv 86$, and

$$
\begin{aligned}
5^{29} &= 5^{16} \times 5^8 \times 5^4 \times 5 \\
&\equiv 86 \times 67 \times 30 \times 5 \\
&\equiv 86 \times 67 \times 150 \\
&\equiv 86 \times 67 \times 31 \\
&\equiv 86 \times 2077 \\
&\equiv 86 \times 54 \\
&\equiv 4644 \equiv 3.
\end{aligned}
$$

where $\equiv$ means $\equiv$ (mod 119) in every case.

RSA can be used to construct a *public key cryptosystem*. Suppose several different people wish to send secret messages to the same recipient. One example is a brokerage house, where the customers want to send their brokers orders to buy or sell; another is military intelligence, where different operatives all report to the same central office. In this case, the recipient could publish the key—the product $pq$ and the power $r$. This is enough information for a member of the public to encrypt a message. However, when the ciphertext is received, the least common multiple of $p - 1$ and $q - 1$ is needed for decryption. In theory, this can be calculated if you know $pq$; you just break this number down into its factors $p$ and $q$, and then you know $p - 1$ and $q - 1$. For example, it takes only a fraction of a second for a computer to find that $119 = 7 \times 17$. But breaking a large number down into factors is a very hard computation. It is possible to choose primes $p$ and $q$ so large that, if you are given only their product, it would take a computer many years to find $p$ and $q$. Your enemy doesn't have years to spend, so only the proper recipient can decrypt the ciphertext.

In practice, you would not transform "CAP" into the sequence 3 1 16. Instead you would run the numbers together, putting in a 0 so that A is written as 01 and C becomes 03 (every letter becomes two numbers), and encrypt the number 030116. In general one could make the whole message into one big number (either ignoring spaces between words, or putting 27 for a space), break it up into a sequence of ten-digit numbers, and encrypt these numbers.

# Multiple Choice Questions 11

**1.** In the scytalee method, what is the *key*?

   A.  the staff                     B.  the diameter of the staff

   C.  the strip of parchment         D.  the plaintext

**2.** How would Caesar encode *"send troops"* (if he spoke English)?

   A.  PBKA QOLLMP             B.  TFOE USPPWT

   C.  VHQG WURRSV           D.  RDMC SQNNOR

**3.** A substitution code produces the ciphertext *KLKEV QOLLMPAOOFWER*. Assuming the message was in simple English, what was the keyword?

   A.  PROW       B.  AZURE       C.  BOXCAR    D.  OXEN

**4.** Which of the following would be a good keyword for a substitution code?

   A.  UP          B.  COPPER       C.  RAYON      D.  FACED

**5.** What is $144(\bmod\, 7)$?

   A.  0          B.  20         C.  3          D.  4

**6.** What is $3^3 - 2(\bmod\, 4)$?

   A.  0          B.  1          C.  2          D.  3

**7.** Use the RSA scheme with $pq = 85$ and $s = 3$ to decode the message "13."

   A.  39         B.  46        C.  63          D.  72

**8.** For the RSA scheme with $p = 13$ and $q = 23$, which of the following could be chosen as a value for $r$?

   A.  6          B.  5          C.  9          D.  33

**9.** For the RSA scheme with $m = 8$ and $r = 5$, what is the value of $s$?

   A.  1          B.  2          C.  3          D.  5

# Exercises 11

**1.** What is the difference in meaning between the words *encode* and *encrypt*?
*In Exercises 2–7, a scytalee message is given. Decrypt it.*

**2.** PSNRMLEDEIESMAEAEORS

**3.** WALOANLFKDTFESHEUMEEPEC

**4.** WRLHSEETRNASRCWOIEMTREYEWIKATISSNFRGIUS

**5.** HRSOETWMWMIEALMNEAYSRMMCOUH

**6.** TNIOHEHSNFOTEWVTDWSEEHORPRN
IFIAETSSTRTOMEITHRECNAESTRG

**7.** LGOABIAIRFRENPKEHSEGHEAIENIIMTSESNCAGTMACSNEOS
DRUYNRTLYBOEYOYPTFRCHLTLTAAAAOEHLEVCGTESSEKRY.

*In Exercises 8 and 9, how would Caesar encode the message (if he spoke English)?*

**8.** *send troops*                           **9** *the end is near*

*In Exercises 10 and 11, assume Caesar spoke English. What command was he sending?*

**10.** DWWDFN                              **11** UHWUHDW

*In Exercises 12 and 13, encrypt the following messages using an additive cipher with key 7.*

**12.** *the moon has risen*                **13** *seven knights are approaching*

*In Exercises 14 and 15, decrypt the following messages using an additive cipher with key 5.*

**14.** IT STY UFXX LT                    **15** GWJFI FSI HNWHZXJX

**16.** Use a substitution cipher with keyword *DROVE* to encrypt the message "cipher is another word for code."

**17.** Use a substitution cipher with keyword *EXTRA* to encrypt the message "cipher is another word for encrypt."

**18.** If you are working modulo 5, what are the results of the calculations

   (i)  $3 + 5$;                    (ii)  $4 \times 3$;                    (iii)  $8 - 2$?

**19.** Determine the value of $x$ so that $5x = 1 \bmod 17$.

**20.** For the RSA scheme with $p = 5, q = 11$, what are the possible $r$ values if $r < 10$?

**21.** For the RSA scheme with $p = 5, q = 11, r = 3$, encode the plaintext "23."

**22.** In using the RSA scheme for large numbers, why do we write A as "01," not just "1"?

# Chapter 12
# Voting Systems

There are many situations that call for a group decision. At one extreme, three of us might be trying to decide where to go for dinner. At the other end of things, millions of people often need to decide which individual, or which political party, will lead their country. Very often we decide by *voting*. But what is the best—fairest, most representative—voting system? This is more complicated, and less obvious, than you might think.

When there are just two candidates for a post, it is all very simple: just vote for the person you prefer. But, as soon as there are more, confusion arises. Remember the 2000 Presidential election. Many people believe the presence of Ralph Nader on the 2000 ballot affected the results.

And this is a very straightforward case. Only one person was to be elected—what we will call a *simple* ballot. One can also have a *complex* election, where more than one candidate is to be elected. For example, if a club is to have a president, a secretary and three committee members, there may be two simple elections—for president and secretary—plus a complex election for the other three.

Other sorts of election are possible; for example, the number to be elected might not be fixed. Some experts have proposed elections where points are allocated to the candidates and the high scorer wins. And sporting contests, where points are awarded not by electors (the judges) but based on the players' performance, are very similar to elections of that kind.

In this chapter we shall look at some of the systems that have been suggested for simple ballots and are used in various places. And in the following chapter we shall cover complex ballots and discuss some of the complications that arise in voting.

W.D. Wallis, *Mathematics in the Real World*, DOI 10.1007/978-1-4614-8529-2_12,
© Springer Science+Business Media New York 2013

## 12.1  Simple Voting

Suppose two people are running for an office. After each person makes his or her one vote, the person who gets more than half the votes wins. This is called the *majority* or *absolute majority* method. Ties are possible—ties are possible in any electoral system—but apart from this the absolute majority method always produces a result.

If there are three or more candidates, the majority method is not so good; there may quite easily be no winner. Several schemes have been devised that allow a candidate with an absolute majority to be elected, and try to find a good approximation when there is no "absolute" winner. These are called *majoritarian* or *plurality* systems.

The first generalization is the *plurality* method, often called "first-past-the-post" voting. (It is also called the *simple majority* method, although a majority is not always involved.) Each voter makes one vote, and the person who receives the most votes wins. For example, if there were three candidates, $A$, $B$ and $C$, and 100 voters, the absolute majority method requires 51 votes for a winner. If $A$ received 40 votes and $B$ and $C$ each got 30, there would be no winner under absolute majority; under plurality, $A$ would be elected.

The problem with the plurality method is that the winner might be very unpopular with a majority of voters. In our example, suppose all the supporters of $B$ and $C$ thought that both these candidates were better-qualified than $A$. Then the plurality method results in the election of the candidate that the majority thought was the worst possible choice. This problem is magnified if there are more candidates; even if there are only four or five candidates, people often think the plurality method elects the wrong person.

To overcome this difficulty in countries with only two major political parties, it is common for each party to endorse only one candidate. For example, in the United States, if there are two or more members of the Republican party who wish to run for some office, a preliminary election, called a *primary election*, is held, and party members vote on the proposed candidates; the one that receives the most votes is nominated by the party, and usually the others do not stand for election. This election would be called the *Republican primary*. There usually will also be a Democratic primary, and sometimes other parties run primaries. However, this method will not solve the problems if there are several major parties, or if the post for which the election is held is not a political one.

## 12.2  Sequential Voting

Another technique used to avoid the problems of the plurality method is *sequential voting*. In this scheme a vote is taken, as a consequence a new set of candidates is selected; then a new vote is taken. The aim is to reduce the set of candidates to a manageable size—often to size two. The original election is again called a *primary*, but in this case all the candidates run in the primary election, not just those in one party.

For example, when the President of France is to be elected, there is a *primary election* for all the candidates. Each voter can select one candidate, and the numbers of votes for each candidate are counted. Later there is another election; there are two candidates, the two candidates who received the most votes in the primary. This second (*runoff*) election is decided by the majority method. A similar method is used in electing the President of the Ukraine, and in a number of other situations.

We shall refer to this as the *runoff method* or *plurality runoff method*. The two top candidates are decided by plurality vote; all other candidates are eliminated; then a majority vote is taken.

In the real world, the preference list is not fixed. There is usually a delay after the primary, and more campaigning takes place. As a result of this the preference lists may change. But for simplicity's sake we shall ignore this for the moment, and assume that every voter has an order of preference between the candidates that remains fixed throughout the voting process. We define the *preference profile* of an election to be the set of all the voters' preference lists. This can conveniently be written in a table. For example, say there are three candidates, $A, B, C$, and suppose:

> 5 voters like $A$ best, then $B$, then $C$;
> 7 voters like $B$ best, then $A$, then $C$;
> 4 voters like $A$ best, then $C$, then $B$;
> 3 voters like $C$ best, then $B$, then $A$;
> no voters like $B$, then $C$, then $A$;
> no voters like $C$, then $B$, then $A$.

We can represent this as

| 5 | 7 | 4 | 3 |
|---|---|---|---|
| $A$ | $B$ | $A$ | $C$ |
| $B$ | $A$ | $C$ | $B$ |
| $C$ | $C$ | $B$ | $A$ |

(We could also write

| 5 | 7 | 4 | 3 | 0 | 0 |
|---|---|---|---|---|---|
| $A$ | $B$ | $A$ | $C$ | $B$ | $C$ |
| $B$ | $A$ | $C$ | $B$ | $C$ | $A$ |
| $C$ | $C$ | $B$ | $A$ | $A$ | $B$ |

but we shall usually omit zero columns.)

In the primary election, each elector votes for the candidate he or she likes best. In the example, $A$ would receive nine votes—five from those with preference list $ABC$, and four from those with list $ACB$. In general, a candidate receives the votes of those who put that candidate first in the preference list. Then the two candidates with the most votes go on to the main election. To work out the votes, you could delete all the other candidates from the electors' lists, so that everybody has a list with just two members. And remember, in the real world, some electors will change their preferences before the runoff election.

**Sample Problem 12.1** *Suppose the preference profile of an election is*

| 5 | 7 | 4 | 3 |
|---|---|---|---|
| A | B | A | C |
| B | A | C | B |
| C | C | B | A |

*What is the result of the election in the following cases?*

(i) *The majority method is used.*
(ii) *The plurality method is used.*
(iii) *The runoff method is used.*

**Solution.** *A* receives nine votes, *B* receives seven, *C* receives three. So (i) there is no majority winner (as there are 19 voters, 10 votes would be needed), and (ii) *A* is the plurality winner. In a primary election, *A* and *B* are selected to contest the runoff. For the runoff, *C* is deleted, and the preference profile becomes

| 5 | 7 | 4 | 3 |
|---|---|---|---|
| A | B | A | B |
| B | A | B | A |

,

or (combining columns with the same preference list)

| 9 | 10 |
|---|---|
| A | B |
| B | A |

.

So *B* wins the runoff.

**Your Turn.** Repeat this question for an election with preference profile

| 7 | 5 | 8 | 3 | 4 |
|---|---|---|---|---|
| A | A | B | B | C |
| B | C | A | C | B |
| C | B | C | A | A |

.

Modifications of this method are also used. When the city of Carbondale, Illinois, elects its Mayor, there is a primary election when there are more than four Mayoral candidates. In the final election, the candidates are the four original candidates who received the most primary votes, and the plurality method is used.

## 12.3 Preferential Voting: The Hare Method

We introduced preference lists as a way of representing a voter's thoughts about the various candidates; they were not actual, physical lists. However, a number of methods have been devised that require a voter to present a preference list. These methods are known collectively as *preferential voting* or *instant runoff voting*. Some methods require the voter to list all candidates; others allow a partial list.

The *Hare method* or *alternative vote* system was invented by the English lawyer Sir Thomas Hare in 1859. It is most useful in its more general form, for situations where several representatives are to be elected at once (see the next chapter). For the moment we look at the simpler version.

The Hare method *requires* each voter to provide a preference list at the election. This list is called a *ballot*. The candidate with the fewest first place votes is eliminated. Then the votes are tabulated again as if there were one fewer candidate, and again the one with the fewest first place votes in this new election is eliminated. When only two remain, the winner is decided by a majority vote.

**Sample Problem 12.2**  *Suppose there are four candidates for a position, and 24 voters whose preference profile is:*

| 5 | 7 | 4 | 3 | 3 | 2 |
|---|---|---|---|---|---|
| A | B | A | C | D | D |
| C | C | D | D | A | C |
| D | D | C | B | B | B |
| B | A | B | A | C | A |

*Who would win using the following electoral systems?*

 (i) *The plurality method.*
 (ii) *The runoff method.*
 (iii) *The Hare method.*

**Solution.**

 (i) The votes for $A$, $B$, $C$ and $D$ are 9, 7, 3 and 5 respectively, so $A$ would win under plurality voting.
 (ii) Under the runoff method there is a tie. $A$ and $B$ are retained, and the new preference profile is:

| 5 | 7 | 4 | 3 | 3 | 2 |
|---|---|---|---|---|---|
| A | B | A | B | A | B |
| B | A | B | A | B | A |

giving 12 votes to each candidate.
 (iii) In the Hare method we first eliminate $C$, who received the fewest first-place votes, obtaining

| 5 | 7 | 4 | 3 | 3 | 2 |
|---|---|---|---|---|---|
| A | B | A | D | D | D |
| B | D | D | B | A | B |
| D | A | B | A | B | A |

Now we eliminate $B$:

| 5 | 7 | 4 | 3 | 3 | 2 |
|---|---|---|---|---|---|
| $A$ | $D$ | $A$ | $D$ | $D$ | $D$ |
| $D$ | $A$ | $D$ | $A$ | $A$ | $A$ |

.

So $D$ wins $15 - 9$.

**Your Turn.** Repeat the above question for the initial preference profile

| 6 | 7 | 7 | 7 | 2 | 7 | 5 | 2 |
|---|---|---|---|---|---|---|---|
| $A$ | $B$ | $A$ | $C$ | $D$ | $D$ | $B$ | $D$ |
| $C$ | $D$ | $D$ | $D$ | $C$ | $C$ | $D$ | $B$ |
| $D$ | $A$ | $C$ | $A$ | $A$ | $B$ | $C$ | $C$ |
| $B$ | $C$ | $B$ | $B$ | $B$ | $A$ | $A$ | $A$ |

.

One modification of this method is to allow voters to cast votes only for the candidates of whom they approve. If there are five candidates, and you think $A$ is best, $B$ second, $C$ third, but do not think either $D$ or $E$ is worthy of election, you simply vote $A, B, C$. If $A$, $B$ and $C$ are all eliminated, your vote is deleted, and the total number of votes cast is reduced accordingly. We shall refer to these generalized Hare systems as *instant runoff systems*.

In the above Sample Problem, suppose the seven voters represented by the second column all decided they did not wish to see $A$ or $D$ elected, while those represented by the fifth column did not like $C$. Then the preference profile is:

| 5 | 7 | 4 | 3 | 3 | 2 |
|---|---|---|---|---|---|
| $A$ | $B$ | $A$ | $C$ | $D$ | $D$ |
| $C$ | $C$ | $D$ | $D$ | $A$ | $C$ |
| $D$ | | $C$ | $B$ | $B$ | $B$ |
| $B$ | | $B$ | $A$ | | $A$ |

.

Again we first eliminate $C$,

| 5 | 7 | 4 | 3 | 3 | 2 |
|---|---|---|---|---|---|
| $A$ | $B$ | $A$ | $D$ | $D$ | $D$ |
| $B$ | | $D$ | $B$ | $A$ | $B$ |
| $D$ | | $B$ | $A$ | $B$ | $A$ |

.

Now we eliminate $B$:

| 5 | 4 | 3 | 3 | 2 |
|---|---|---|---|---|
| $A$ | $A$ | $D$ | $D$ | $D$ |
| $D$ | $D$ | $A$ | $A$ | $A$ |

.

and $A$ wins $9 - 8$.

Although nine votes is not a majority of the original 24 voters, it may count as a majority for this purpose. In some cases a *quota* is declared: a number is decided, and if neither of the last two candidates remaining achieves that many votes then the election is declared void. So, if there was a quota of 10 votes, the election would have to be held again.

A complication that can occur in real life is that a voter might not prefer one candidate over another: ties could occur in the voter's preferences. In practice, electoral systems that use preference lists do not allow ties, so that the voter must make a (possibly arbitrary) choice between the tied candidates. For simplicity, we shall assume that there are no ties in preference profiles.

## 12.4 The Condorcet Method

A multiple use of runoff elections was discussed by Marie Jean Antoine Nicolas Caritat, Marquis de Condorcet, an eighteenth Century French mathematician and political theorist. (Similar ideas were proposed by Ramon Llull as long ago as 1299.)

Suppose we simultaneously conduct all the "runoff" elections among our candidates. For example, in the election discussed in Sample Problem 12.2, there are six runoffs: $A$ versus $B$, $A$ versus $C$, $A$ versus $D$, $B$ versus $C$, $B$ versus $D$, and $C$ versus $D$. If any one candidate wins all his/her runoffs, then surely you would consider that person a winner. We shall call such a candidate a *Condorcet winner*.

In Sample Problem 12.1, we find:

$B$ beats $A$ 10–9,
$A$ beats $C$ 16–3,
$B$ beats $C$ 12–7,

so $B$ is a Condorcet winner. Similarly, in Sample Problem 12.2, $D$ is a Condorcet winner. But even in the simple example

| 5 | 4 | 3 |
|---|---|---|
| A | B | C |
| B | C | A |
| C | A | B |

$A$ beats $B$ 8 – 4, $B$ beats $C$ 9 – 3 and $C$ beats $A$ 7 – 5, so there is no Condorcet winner.

In elections with several candidates, it is very common to have no Condorcet winner, even when there are no ties. This is a serious fault in the Condorcet method.

Condorcet's own solution to this problem is as follows. We shall construct an ordered list of the candidates. Look at all the runoffs and find out which candidate won with the biggest majority. Looking at Sample Problem 12.1 again, the biggest majority was $A$ beat $C$ 16–3. We'll denote this $A \to C$. Then look for the second-biggest, then the third-biggest, and so on, and make a list:

$$A \rightarrow C(16-3), B \rightarrow C(12-7), B \rightarrow A(10-9).$$

Now go through this list and construct a preference order of the candidates. At each step, if $X \rightarrow Y$, then $X$ precedes $Y$ in the preference list, *unless* $Y$ already precedes $X$ in the list. In our example, we must have $A$ before $C$, $B$ before $C$ and $B$ before $A$. The list is $BAC$ and clearly $B$ is the winner.

We shall refer to this solution as *Condorcet's extended method*, to distinguish it from the case where there is a Condorcet winner under the original method. Note that, if there is a Condorcet winner, the same candidate also wins under the extended method.

Let us apply this to the above example

| 5 | 4 | 3 |
|---|---|---|
| A | B | C |
| B | C | A |
| C | A | B |

which has no Condorcet winner. We have, with the larger majorities preceding smaller ones,

$$B \rightarrow C(9-3), A \rightarrow B(8-4), C \rightarrow A(7-5).$$

From $B \rightarrow C$ and $A \rightarrow B$ we get the list $ABC$. Next we see $C \rightarrow A$, but $A$ already precedes $C$, so this result is ignored. The final list is $ABC$, and $A$ is elected, even though a majority of voters would prefer $C$ to $A$.

**Sample Problem 12.3** *Consider the election with preference profile:*

| 7 | 5 | 3 |
|---|---|---|
| A | B | C |
| B | C | A |
| C | A | B |

*Who would win under the Hare method? Is there a Condorcet winner? Who wins under Condorcet's solution method?*
**Solution.** The votes for $A$, $B$ and $C$ are 7, 5 and 3 respectively. Under the Hare method, $C$ is eliminated. The new preference profile is:

| 7 | 5 | 3 |
|---|---|---|
| A | B | A |
| B | A | B |

that is,

| 10 | 5 |
|----|---|
| A  | B |
| B  | A |

So $A$ wins $10-5$. Looking at all three runoffs, we see that $A$ beats $B$ $10-5$, $B$ beats $C$ $12-3$ and $C$ beats $A$ $8-7$, so there is no Condorcet winner. For Condorcet's solution, we see

$$B \to C(12-3), A \to B(10-5), C \to A(7-5).$$

The first two yield the list $ABC$ and the last result is ignored, so $A$ is elected.

## 12.5   Sequential Pairwise Voting

In *sequential pairwise voting* several candidates are paired in successive runoff elections. There is an *agenda* (an ordered list of candidates). For example, if the agenda is $A, B, C, D, \ldots$ then the elections proceed as follows:

1. $A$ against $B$
2. Winner of $AB$ against $C$
3. That winner against $D$

    $\cdots$

Position in the agenda is very important. To see this, consider a four-candidate election with agenda $A, B, C, D$, in which all four candidates are equally likely to win. If repeated trials are made then we would expect the following results:

- $A$ wins first runoff in half the cases
- $A$ wins second runoff in half those cases—a quarter overall
- $A$ wins the third runoff in half those cases—one-eighth overall.

So $A$ has a 1 in 8 chance of winning. $B$ also has a 1 in 8 chance. However, $C$ has a 1 in 4 chance, and $D$ has a 1 in 2 chance. In this case, being later in the list is very beneficial.

Rather than elections, this model is often used for sporting tournaments (the result of match is used instead of the result of a runoff election). One often sees playoff rules like:

(i) Second and third placegetters in preliminary competition play each other ("the playoff")
(ii) the winner of playoff meets the leader from the preliminaries.

In this case it is reasonable that the preliminary leader should get an advantage. However, when the model is used in voting situations, it is very subject to manipulation.

## 12.6   Point Methods

Pointscore methods have often been used in sporting contests. For example, they are commonly used in track meets and in motor racing. When the Olympic Games are being held, many newspapers publish informal medal tallies to rank the performance of the competing nations—the usual method is to allocate three points for a gold medal, two for a silver, one for a bronze, and then add.

In general, a fixed number of points are given for first, second, and so on. The points are totaled, and the candidate with the most points wins. If there are $n$ competitors, a common scheme is to allocate $n$ points to first, $n-1$ to second, ..., or equivalently $n-1$ to first, $n-2$ to second, .... This case, where the points go in uniform steps, is called a *Borda count*.

One often see scales like 5, 3, 2, 1, where the winner gets a bonus, or 3, 2, 1, 0, 0, ... (that is, all below a certain point are equal). Sometimes more complicated schemes are used; for example, in the Indy Racing League, the following system has been used:

| | | | | | |
|---|---|---|---|---|---|
| 1st gets | 20 | 5th gets | 10 | 9th gets | 4 |
| 2nd gets | 16 | 6th gets | 8 | 10th gets | 3 |
| 3rd gets | 14 | 7th gets | 6 | 11th gets | 2 |
| 4th gets | 12 | 8th gets | 5 | 12th gets | 1 |

Fastest qualifier gets one point.
Leader of most laps gets one point.

Pointscore methods are occasionally employed for elections, most often for small examples such as selection of the best applicant for a job.

Sometimes the result depends on the point scheme chosen.

**Sample Problem 12.4** *What is the result of an election with preference table*

| 5 | 7 | 4 | 3 |
|---|---|---|---|
| A | B | A | C |
| C | C | C | B |
| B | A | B | A |

*if a 3, 2, 1 count is used? What is the result if a 4, 2, 1 count is used?*
**Solution.** With a 3, 2, 1 count the totals are $A : 37, B : 36, C : 41$, so $C$ wins. With a 4, 2, 1 count the totals are $A : 46, B : 43, C : 44$, and $A$ wins.

# Multiple Choice Questions 12

1. A majority election is held, with 200 electors. How many votes are required to win?
   A.  99                 B.  100                 C.  101
   D.  None of the above answers is correct
2. In the election of a committee chairman, there are 27 electors and 3 candidates. How many votes are required to win a majority election?
   A.  9            B.  10            C.  14            D.  27
   *Questions 3–7 refer to the preference profile*

   | 4 | 3 | 3 | 3 |
   |---|---|---|---|
   | B | C | D | C |
   | D | A | A | B |
   | C | D | B | A |
   | A | B | C | D |

3. Who wins if it is a plurality election?
   A.  A            B.  B                        C.  C
   D.  D            E.  There is no winner
4. Who wins if it is a runoff election?
   A.  A            B.  B                        C.  C
   D.  D            E.  There is no winner
5. Who wins if a $(3,2,1,0)$ Borda count is used?
   A.  A            B.  B                        C.  C
   D.  D            E.  There is no winner
6. Who wins under the Hare system?
   A.  A            B.  B                        C.  C
   D.  D            E.  There is no winner
7. Who wins if it is a majority election?
   A.  A            B.  B                        C.  C
   D.  D            E.  There is no winner
   *Questions 8 and 9 refer to the preference profile*

   | 6 | 5 | 5 | 2 |
   |---|---|---|---|
   | A | B | C | D |
   | D | D | B | B |
   | C | C | A | A |
   | B | A | D | C |

8. Who wins under the Hare voting system?
   A.  A            B.  B            C.  C            D.  D
9. Who wins under sequential pairwise voting, with agenda $A, B, C, D$?
   A.  A            B.  B            C.  C            D.  D

**10.** Consider the following statements.

(i) If $X$ would win a plurality election, then $X$ would win under Condorcet.

(ii) If $X$ would win a majority election, then $X$ would win under Condorcet.

Which of the statements is always true?

A. (i), not (ii)  B. (ii), not (i)

C. Both statements  D. Neither statement

## Exercises 12

1. How many votes are needed for a majority winner if there are 35 voters?

2. In how many ways can a voter rank five candidates assuming ties are not allowed?

3. Twenty-eight electors vote between candidates $A$, $B$ and $C$. Their votes are 4 for $A$, 15 for $B$ and 9 for $C$. What is the result under the majority method? What is the result under the plurality method?

4. Seventy-five electors vote between candidates $A$, $B$ and $C$. There were 25 votes for $A$ and 28 votes for $B$. How many votes did $C$ receive? What is the result under the majority method? What is the result under the plurality method?

5. First-year students in a University vote for Class president. There are three candidates, Smith, Jones, and Brown. The preference table is

| 25 | 27 | 14 | 22 | 35 |
|----|----|----|----|----|
| S | S | J | J | B |
| J | B | S | B | J |
| B | J | B | S | S |

(i) How many students voted?

(ii) How many first place votes did each candidate receive?

(iii) Who, if anybody, would win under the plurality method?

(iv) Who, if anybody, would win under the majority method?

6. At the Academy Awards there are three nominees for Best Actor: Arthur Andrews, Bob Brown and Clive Carter. The preference table is

| 123 | 101 | 442 | 212 | 315 |
|-----|-----|-----|-----|-----|
| A | A | B | C | C |
| B | C | A | A | B |
| C | B | C | B | A |

(i) How many actors voted?

(ii) How many first place votes did each candidate receive?

(iii) Who, if anybody, would win under the plurality method?

(iv) Who, if anybody, would win under the majority method?

(v) Who would win under the runoff method?

**7.** In addition to plurality and the runoff method, two other techniques have been devised for cases when there is no majority winner. Both assume the full preference lists are known.

(A) The winner is the candidate with the fewest last-place votes.

(B) A runoff is held between the two candidates with the fewest last-place votes.

What are the results of using these two methods:

(i) using the data of Exercise 5?

(ii) using the data of Exercise 6?

**8.** Thirty board members must vote on five candidates: $X$, $Y$, $Z$, $U$, and $V$. Their preference rankings are summarized in the table below. Find the winner using sequential pairwise voting with the agenda $X,Y,Z,U,V$.

| 12 | 10 | 8 |
|----|----|---|
| X | Y | Z |
| U | U | U |
| Y | Z | X |
| Z | X | V |
| V | V | Y |

**9.** Nineteen sports announcers rank the final four in a basketball tournament. The teams are Creighton ($C$), Drake ($D$), Evansville ($E$), and Southern Illinois ($S$). Their preference table is

| 9 | 3 | 2 | 5 |
|---|---|---|---|
| S | C | E | C |
| C | S | D | E |
| E | D | C | D |
| D | E | S | S |

Which team will they choose as favorite, if they use the following methods?

(i) Plurality.

(ii) A Borda count.

(iii) The Hare method.

**10.** Twenty-one electors wish to choose between four candidates $A,B,C,D$. The preference table is

| 8 | 9 | 4 |
|---|---|---|
| C | B | A |
| B | C | C |
| A | D | B |
| D | A | D |

  (i)  Who is elected under the plurality method?
  (ii)  Who is elected under the Hare method?
11. Fifteen voters must choose one representative from four candidates: $A$, $B$, $C$, and $D$. Their preference profile is

| 6 | 3 | 4 | 2 |
|---|---|---|---|
| A | C | C | B |
| B | B | D | A |
| C | D | A | C |
| D | A | B | D |

What is the result of:
  (i)  A majority rule election?
  (ii)  A plurality election?
  (iii)  An election using the Hare method?
12. Twenty-one electors must choose between five candidates: $V$, $W$, $X$, $Y$, and $Z$. Their preference rankings are:

| 4 | 3 | 6 | 3 | 2 | 3 |
|---|---|---|---|---|---|
| V | V | X | Y | Y | W |
| W | Y | Z | X | X | Z |
| X | W | Y | W | W | V |
| Y | X | V | Z | V | X |
| Z | Z | W | V | Z | Y |

If there is no majority winner, all candidates with fewer than 20% of the first-place votes (that is, those with fewer than 5) will be eliminated, and the preferences adjusted accordingly. If there is still no winner, a runoff is held. Which candidates are eliminated? What is the final result?
13. Given the following preference table, who would win under plurality voting? Who would win in a runoff?

| 6 | 3 | 4 | 2 | 1 |
|---|---|---|---|---|
| A | C | C | B | E |
| E | B | D | A | A |
| B | E | A | C | B |
| C | D | E | D | C |
| D | A | B | E | D |

Is there a Condorcet winner?

**14.** A club with 36 members wishes to elect its president from four candidates, $A$, $B$, $C$ and $D$. The preference profile is

| 16 | 10 | 8 | 2 |
|----|----|---|---|
| A  | B  | C | B |
| B  | A  | B | A |
| C  | D  | A | C |
| D  | C  | D | D |

.

(i) Who would be elected if the club used plurality voting?

(ii) Who would be elected if the club used the (3, 2, 1, 0) Borda count?

(iii) Who would be elected if the club used a modified Borda count with scores (5, 2, 1, 0)?

(iv) Is there a Condorcet winner? If not, who would win under Condorcet's extended method?

**15.** Eighteen delegates must elect one of four candidates: $A$, $B$, $C$, and $D$. The preference profile is

| 8 | 6 | 4 |
|---|---|---|
| A | B | C |
| B | D | D |
| C | A | A |
| D | C | B |

.

(i) Who would be elected under the Hare method?

(ii) Who would be elected if the delegates used a modified Borda count with scores (2, 1, 0, 0)?

(iii) Is there a Condorcet winner? If not, who would win under Condorcet's extended method?

**16.** Who would win the election of Exercise 5, if a (3, 2, 1) Borda count were used?

**17.** Who would win the election of Exercise 12, if a (4, 3, 2, 1, 0) Borda count were used?

**18.** Here is the preference profile for an election with three candidates, $A$, $B$, $C$. Is there a Condorcet winner? If not, who would win under Condorcet's extended method?

| 8 | 5 | 6 | 4 |
|---|---|---|---|
| A | C | B | B |
| C | A | C | A |
| B | B | A | C |

.

**19.** Fifty voters are to choose one of five candidates. Their preference profile is

| 20 | 10 | 14 | 6 |
|----|----|----|---|
| A  | B  | B  | C |
| C  | A  | A  | D |
| E  | C  | D  | B |
| B  | D  | C  | A |
| D  | E  | E  | E |

What is the result under the following methods?
   (i) Plurality.
   (ii) Runoff.
   (iii) The Hare method.
   (iv) A modified Borda count with scores (5, 3, 2, 1, 0).
   (v) The *reverse* Hare method: after each ballot, the candidate with the most last place votes is eliminated.
Is there a Condorcet winner? If not, who would win under Condorcet's extended method?

**20.** Fifteen committee members are to choose a new treasurer from four candidates: $A, B, C$, and $D$. Their preference profile is

| 7 | 5 | 3 |
|---|---|---|
| A | C | D |
| B | B | C |
| D | A | B |
| C | D | A |

Is there a Condorcet winner? What is the result under:
   (i) Plurality?
   (ii) The Hare method?
   (iii) The extended Condorcet method?
   (iv) The reverse Hare method described in Exercise 19?

**21.** One hundred voters choose between four candidates, $A, B, C, D$. Their preference profile is

| 40 | 32 | 10 | 18 |
|----|----|----|----|
| A  | B  | C  | D  |
| C  | C  | D  | C  |
| B  | A  | A  | B  |
| D  | D  | B  | A  |

What is the result if the Hare method is used? Is there a Condorcet winner?

22. Suppose there are three candidates in an election, where one candidate is to be elected. Is there any difference whether the runoff method or the Hare method is used?

23. Twenty-five electors vote for three candidates, resulting in the preference table

| 8 | 6 | 7 | 4 |
|---|---|---|---|
| X | Y | Z | Y |
| Z | Z | X | X |
| Y | X | Y | Z |

It is decided to use a scoring system where first place gets $n$ points, second gets 2 and third gets 1, where $n$ is some whole number greater than 2. For what ranges of values is $X$ winner? What is the range for $Y$? For $Z$? Does it ever happen that there is no result?

24. Explain why majority rule is a reasonable electoral method in a country with only two political parties, but is not good in a country with four major political parties. (Why is the word "major" important here?)

25. When deciding an election, is it necessary to know the number of voters associated with each preference list, or is it sufficient to know the percentage of voters?

26. Construct a preference profile for four voters and four candidates, such that three voters prefer $W$ to $X$, three prefer $X$ to $Y$, three prefer $Y$ to $Z$, and three prefer $Z$ to $W$.

27. Verify that if an election has a Condorcet winner, then the same candidate also wins under Condorcet's extended method.

# Chapter 13
# More on Voting

In some elections, several candidates are to be elected simultaneously. For example, in Australia, the Senate (the upper house of Parliament) contains 12 representatives from each state, and six are elected at each election. Another example is the election of a club committee; in addition to a president, a treasurer and a secretary, there will often be some at-large members who can all be elected at once. To start this chapter we shall introduce two methods used in multiple elections. We shall then describe the ways in which some voting systems can be manipulated. We conclude by looking at how different systems, while appearing quite fair, can produce wildly different results.

## 13.1  The Generalized Hare Method

The Hare system was modified by Andrew Inglis Clark, who was Attorney-General of Tasmania in the late nineteenth century. This *generalized Hare method*, or *quota method*, is used in many countries, including Ireland, Australia and Malta.

As an example, suppose 240,000 people are to elect five representatives from a larger number of candidates.

Each person makes a *preferential* vote, listing *all* the candidates in order. (In other words, they record all of their preference profiles, not just the first members.) The first name on the list is called the voter's *first preference*, and so on.

It is impossible for six candidates each to get more than 40,000 first preference votes. So any candidate who receives more than 40,000 first-place votes is elected. This number is called the *quota*. If a candidate receives more than 40,000 votes, the remainder of those votes go to voters' second choices, divided proportionally. To illustrate this, suppose $A$ gets 50,000 votes. Of these, 25,000 have $B$ as second choice, 20,000 have $C$ and 5,000 have $D$. Since this is greater than 40,000, $A$ is declared elected. The 10,000 surplus votes are divided in the proportion 25,000:20,000:5,000, or 50% : 40% : 10%. That is, $B$ gets 50% of $A$'s preferences,

W.D. Wallis, *Mathematics in the Real World*, DOI 10.1007/978-1-4614-8529-2_13,
© Springer Science+Business Media New York 2013

because 25,000 is 50% of 50,000; $C$ gets 40%; and $D$ gets 10%. As we said, $A$ has exceeded the quota by 10,000 votes, so we say he has 10,000 preferences to be distributed. $B$ gets 50% of these, so 5,000 more votes are added to $B$'s total. In the same way $C$ gets 4,000 added votes (40%) and $D$ gets 1,000 votes (10%).

Now we check whether $B$, $C$ or $D$ has exceeded the quota. For example, if $B$ previously had 37,000 votes, the new total would be 42,000, exceeding the quota again, so $B$ is declared elected.

If not enough candidates have been elected, and no one has enough votes, the remaining candidate with the *fewest* votes is eliminated. *All* of that candidate's preferences are distributed (in just the same way as in the regular Hare method, so no percentages need to be calculated).

In general, suppose there are $V$ voters and $N$ places to be filled. The quota is

$$Q = \frac{V}{N+1},$$

and a candidate is declared to be elected if his or her number of votes exceed the quota. Notice that, if only one candidate were to be elected, the quota requirement is that a candidate receive a majority of the votes. (Some systems use the formula $\frac{V}{N+1} + 1$, and say a candidate is elected if the quota is equaled or exceeded. This may require one more vote in the case where $\frac{V}{N+1}$ is not an integer.)

It is possible that, after $N - 1$ candidates have been declared elected, the remaining votes are shared between two candidates and each receives exactly $Q$ votes. If this were to happen, a runoff would be held between those two candidates. However, the probability of this happening in a real world example is very small.

In a real example, the whole list of preferences is kept. The process may require a great deal of data; moreover, it may result in complicated numbers, fractions, and so on. Historically this was a serious problem and caused long delays in announcing the results of elections, but it is no longer an issue now that computers are available and voting machines can be adapted to keep all the data.

For example, say a preference profile has the form

| 12 | 9 | 6 | ... | ... |
|----|---|---|-----|-----|
| $A$ | $A$ | $A$ | $B$ | ... |
| $B$ | $B$ | $C$ | ... | ... |
| $C$ | $D$ | $B$ | ... | ... |
| $D$ | $C$ | $D$ | ... | ... |

(where lines of dots represent the votes where $A$ is not in first place; the numbers do not concern us). If the quota is 18, $A$'s surplus is nine. As $A$ received 27 first-place votes, those 27 voters share the nine surplus votes; each receives $\frac{9}{27}$, or one-third, of a vote. So, after $A$ is elected, the 12 voters who are represented in the first column of the profile receive $\frac{1}{3}$ each, a total of four votes; the second and third column voters receive totals of 3 and 2 respectively. $A$ is declared elected, and the process of electing a second candidate proceeds as though the preference profile was

| 4 | 3 | 2 | ... | ... |
|---|---|---|-----|-----|
| $B$ | $B$ | $C$ | $B$ | ... |
| $C$ | $D$ | $B$ | ... | ... |
| $D$ | $C$ | $D$ | ... | ... |

with $A$ deleted from every column. The quota remains 18.

**Sample Problem 13.1** *Say there are five candidates for three positions; the preference table is*

| 6 | 6 | 9 | 6 | 3 | 2 |
|---|---|---|---|---|---|
| $A$ | $A$ | $C$ | $C$ | $E$ | $E$ |
| $B$ | $B$ | $D$ | $D$ | $C$ | $A$ |
| $E$ | $D$ | $E$ | $E$ | $D$ | $B$ |
| $D$ | $E$ | $A$ | $B$ | $A$ | $C$ |
| $C$ | $C$ | $B$ | $A$ | $B$ | $D$ |

*Who will be elected?*

**Solution.** There are 32 voters, so the quota is 8. $A$ gets 12 first place votes and $C$ gets 15, both of which exceed the quota. So $A$ and $C$ are elected.

$A$ has a surplus of 4. The second-place candidate in every case is $B$, but we observe that the votes are divided in proportion 6:6 between $A$'s two lists, so we give two votes to each of $A$'s lists. $C$ is also at the top of two lists, and has a surplus of 7, divided 9:6. This gives surplus allocations of 4.2 and 2.8. After these allocations, the new table is

| 2 | 2 | 4.2 | 2.8 | 3 | 2 |
|---|---|-----|-----|---|---|
| $B$ | $B$ | $D$ | $D$ | $E$ | $E$ |
| $E$ | $D$ | $E$ | $E$ | $D$ | $B$ |
| $D$ | $E$ | $B$ | $B$ | $B$ | $D$ |

$B$ has four votes, $D$ has 7, and $E$ has 5. No one exceeds the quota. (Note that the quota does not change.) So $B$ (who had the fewest votes) is eliminated. We now have

| 2 | 2 | 4.2 | 2.8 | 3 | 2 |
|---|---|-----|-----|---|---|
| $E$ | $D$ | $D$ | $D$ | $E$ | $E$ |
| $D$ | $E$ | $E$ | $E$ | $D$ | $D$ |

$D$ now has 9 votes and $E$ has 7, so $D$ is elected. In total, $A$, $C$ and $D$ are elected.

**Your Turn.** Repeat the above problem for preference profile

| 5 | 5 | 7 | 8 | 5 | 2 |
|---|---|---|---|---|---|
| A | A | B | C | D | E |
| B | E | D | D | C | A |
| E | B | E | E | E | B |
| D | D | A | B | A | C |
| C | C | C | A | B | D |

## 13.2  Approval Voting

Approval voting was first used in cases where the number of people to be elected was not fixed. For example, if a committee wants to co-opt a few more members to help organize a function, they might add all those who they think would make a positive contribution, not a fixed number. In that case a member might vote for every suitable candidate, casting as many votes as she wishes; all the votes are considered equal. The number of votes received by a candidate is called his *approval rating*. Candidates with an approval rating of at least 50% (or 60%, or some other agreed figure) are elected.

Sometimes there is a strict requirement that a certain number be elected, or a minimum or maximum is imposed. Ties are more common in approval voting than they are in some other systems, and some sort of runoff procedure is often necessary. The following example illustrates these ideas.

**Sample Problem 13.2** *Ten board members vote on eight candidates by approval. The candidates are A,B,C,D,E,F,G,H, and the board members are q,r,s,t,u, v,w,x,y,z. They vote as follows ( × represents approval).*

|   | q | r | s | t | u | v | w | x | y | z |
|---|---|---|---|---|---|---|---|---|---|---|
| A | × | × | × |   | × |   | × | × |   | × |
| B |   | × | × | × | × | × | × | × | × |   |
| C |   |   | × |   |   |   |   | × |   |   |
| D | × |   | × | × | × | × | × | × |   | × |
| E | × | × | × | × | × |   |   | × |   |   |
| F | × | × | × | × | × | × | × |   | × | × |
| G | × |   | × | × |   |   | × |   | × |   |
| H |   | × |   | × | × | × |   |   |   | × |

(i) *Which candidate is chosen if just one is to be elected?*

(ii) *Which candidates are chosen if the top four are to be elected?*

*(iii) Which candidates are chosen if the top two are to be elected?*

*(iv) Which candidate or candidates are chosen if at most four are to be elected and 80% approval is required?*

**Solution.** Here is a summary of the votes received:

$$A - 7 \quad B - 8 \quad C - 2 \quad D - 8 \quad E - 6 \quad F - 9 \quad G - 5 \quad H - 5.$$

(i) $F$, who received the most votes.

(ii) $A, B, D, F$.

(iii) $F$ is elected; there will need to be a runoff election between $B$ and $D$.

(iv) $B$, $D$ and $F$ are elected.

The array in the example is called an *approval table*; we shall always denote approval with a cross in such tables.

Approval voting is particularly useful for situations like the selection of new employees. In those circumstances there is usually a minimum requirement, and further applications may be called if not enough good applicants are available.

## 13.3 Manipulating the Vote

The term *strategic voting* means voting in a way that does not represent your actual preferences, in order to change the result of the election. We would call the resulting ballot *insincere*.

Suppose your favorite is candidate $X$. (We will call you an $X$ *supporter*.) Then $X$ would normally appear at the top of your preference list. But sometimes you can achieve $X$'s election by voting for another candidate in first place! This is most common in runoff situations; you can ensure that your candidate does not have to face a difficult opponent. The following example illustrates this.

**Sample Problem 13.3** *A runoff election has preference profile*

| 6 | 2 | 7 | 5 | 4 |
|---|---|---|---|---|
| A | C | C | B | D |
| D | B | A | A | A |
| C | D | D | C | B |
| B | A | B | D | C |

*Show that the supporters of C can change the result so that their candidate wins, by the two voters in the second column changing their ballots by demoting their candidate.*

**Solution.** Initially the first-place votes are $A$–6, $B$–5, $C$–9, $D$–4, so runoff election will be between $A$ and $C$, and $A$ wins 15–9. The revised profile is

| 6 | 2 | 7 | 5 | 4 |
|---|---|---|---|---|
| A | B | C | B | D |
| D | C | A | A | A |
| C | D | D | C | B |
| B | A | B | D | C |

,

the first-place votes are $A$–6, $B$–7, $C$–7, $D$–4, so the runoff election is between $B$ and $C$, and $C$ wins 13–11.

Even when you cannot ensure victory for your favorite candidate, you may still be able to obtain a preferable result. For example, suppose you support candidate $X$; you think candidate $Y$ is acceptable, but hate candidate $Z$. Even if insincere voting cannot ensure victory for candidate $X$, you may be able to swing the election to $Y$ rather than $Z$.

**Sample Problem 13.4** *An election with four candidates and seven voters is to be decided by the Hare system. The preference profile is*

| 2 | 1 | 2 | 1 | 2 |
|---|---|---|---|---|
| B | B | D | C | C |
| A | D | C | A | D |
| D | A | B | B | A |
| C | C | A | D | B |

.

*Show that one of the two voters with $B,A,D,C$ can change the outcome to a more favorable one by insincere voting.*
**Solution.** First consider the result of sincere voting. Initially $A$ is eliminated, having no first-place votes:

| 2 | 1 | 2 | 1 | 2 |
|---|---|---|---|---|
| B | B | D | C | C |
| D | D | C | B | D |
| C | C | B | D | B |

.

Next $D$ is eliminated, leaving

| 2 | 1 | 2 | 1 | 2 |
|---|---|---|---|---|
| B | B | C | C | C |
| C | C | B | B | B |

.

The winner is $C$.

Now suppose one voter changes his ballot from $B,A,D,C$ to $D,A,B,C$. The profile is

| 1 | 1 | 1 | 2 | 1 | 2 |
|---|---|---|---|---|---|
| B | D | B | D | C | C |
| A | A | D | C | A | D |
| D | B | A | B | B | A |
| C | C | C | A | D | B |

.

Again *A* is first eliminated, leaving:

| 1 | 1 | 1 | 2 | 1 | 2 |
|---|---|---|---|---|---|
| B | D | B | D | C | C |
| D | B | D | C | B | D |
| C | C | C | B | D | B |

.

In the next round, *B* is eliminated.

| 1 | 1 | 1 | 2 | 1 | 2 |
|---|---|---|---|---|---|
| D | D | D | D | C | C |
| C | C | C | C | D | D |

;

The winner is *D*. This is a preferable outcome for the voter who switched.

**Your Turn.** Consider a Hare system election with preference profile

| 8 | 6 | 5 |
|---|---|---|
| P | Q | R |
| Q | R | S |
| R | P | P |
| S | S | Q |

.

Show that *P* would win this election. Show that if one of the six supporters of *Q* changes her vote, she could ensure that *R* wins, even though a majority of voters still prefer *Q* to *R*.

Of course, you do not always know exactly how the votes will go. Strategic voting is usually based on assumptions about the election.

## 13.4  Amendments

We now consider an important example of manipulation of sequential pairwise voting. Suppose three voters on City Council have to decide whether to add a new sales tax. Initially

- *A* prefers the tax
- *B* prefers the tax
- *C* prefers no tax

so a tax will be introduced.

However, let's assume *A* hates income taxes and will never vote for one. On the other hand, *B* prefers income tax to sales tax. Suppose *C* moves an amendment to change the tax to an income tax.

We now have:
Original motion: that a city sales tax of 5% be introduced.
Amendment (moved by C): change "sales tax of 5%" to "income tax of 2%."
(We'll assume the 2% income tax will provide the same total as the 5% sales tax.)

In the vote on the amendment, both *B* and *C* will vote in favor, with *A* against, so the amendment is carried. So the motion becomes: a city income tax of 2% shall be introduced. In the vote on the new motion, both *A* and *C* are against, while *B* votes in favor; so the motion is lost and there is no tax.

## 13.5   Five Candidates, Five Winners

We conclude by showing that even fair elections can produce unexpected results.

Consider a political party convention at which five different voting schemes are adopted. Assume that there are 110 delegates to this national convention, at which five of the party members, denoted by *A*, *B*, *C*, *D*, and *E*, have been nominated as the party's presidential candidate. Each delegate must rank all five candidates according to his or her choice. Although there are $5! =: 5 \times 4 \times 3 \times 2 \times 1 = 120$ possible rankings, many fewer will appear in practice because electors typically split into blocs with similar rankings. Let's assume that our 110 delegates submit only six different preference lists, as indicated in the following preference profile:

Number of delegates

|                | 36 | 24 | 20 | 18 | 8 | 4 |
|----------------|----|----|----|----|---|---|
| First choice   | A  | B  | C  | D  | E | E |
| Second choice  | D  | E  | B  | C  | B | C |
| Third choice   | E  | D  | E  | E  | D | D |
| Fourth choice  | C  | C  | D  | B  | C | B |
| Fifth choice   | B  | A  | A  | A  | A | A |

The 36 delegates who most favor nominee $A$ rank $D$ second, $E$ third, $C$ fourth, and $B$ fifth. Although $A$ has the most first-place votes, he is actually ranked last by the other 74 delegates. The 12 electors who most favor nominee $E$ split into two subgroups of 8 and 4 because they differ between $B$ and $C$ on their second and fourth rankings.

We shall assume that our delegates must stick to these preference schedules throughout the following five voting agendas. That is, we will not allow any delegate to switch preference ordering in order to vote in a more strategic manner or because of new campaigning.

We report the results when six popular voting methods are used. There are six different results.

1. **Majority.** As one might expect with five candidates, there is no majority winner.
2. **Plurality.** If the party were to elect its candidate by a simple plurality, nominee $A$ would win with 36 first-place votes, in spite of the fact that $A$ was favored by less than one-third of the electorate and was ranked dead last by the other 74 delegates.
3. **Runoff.** On the other hand, if the party decided that a runoff election should be held between the top two contenders ($A$ and $B$), who together received a majority of the first-place votes in the initial plurality ballot, then candidate $B$ outranks $A$ on 74 of the 110 preference schedules and is declared the winner in the runoff.
4. **Hare Method.** Suppose the Hare method is used: a sequence of ballots is held, and eliminating at each stage the nominee with the fewest first-place votes. The last to survive this process becomes the winning candidate. In our example $E$, with only 12 first-place votes, is eliminated in the first round. $E$ can then be deleted from the preference profile, and all 110 delegates will vote again on successive votes. On the second ballot, the 12 delegates who most favored $E$ earlier now vote for their second choices, that is, 8 for $B$ and 4 for $C$; the number of first-place votes for the four remaining nominees is

| $A$ | $B$ | $C$ | $D$ |
|-----|-----|-----|-----|
| 36  | 32  | 24  | 18  |

Thus, $D$ is eliminated. On the third ballot the 18 first-place votes for $D$ are reassigned to $C$, their second choice, giving

| $A$ | $B$ | $C$ |
|-----|-----|-----|
| 36  | 32  | 42  |

Now $B$ is eliminated. On the final round, 74 of the 110 delegates favor $C$ over $A$, and therefore $C$ wins.

5. **Borda Count.** Given that they have the complete preference schedule for each delegate, the party might choose to use a straight Borda count to pick the winner.

They assign five points to each first-place vote, four points for each second, three points for a third, two points for a fourth, and one point for a fifth. The scores are:

$A$: $254 = (5)(36) + (4)(0) + (3)(0) + (2)(0) + (1)(24 + 20 + 18 + 8 + 4)$
$B$: $312 = (5)(24) + (4)(20 + 8) + (3)(0) + (2)(18 + 4) + (1)(36)$
$C$: $324 = (5)(20) + (4)(18 + 4) + (3)(0) + (2)(36 + 24 + 8) + (1)(0)$
$D$: $382 = (5)(18) + (4)(36) + (3)(24 + 8 + 4) + (2)(20) + (1)(0)$
$E$: $378 = (5)(8 + 4) + (4)(24) + (3)(36 + 20 + 18) + (2)(0) + (1)(0)$

The highest total score of 382 is achieved by $D$, who then wins. $A$ has the lowest score (254) and $B$ the second lowest (312).

6. **Condorcet.** With five candidates, there is often no Condorcet winner. However, when we make the head-to-head comparisons, we see that $E$ wins out over:

- $A$ by a vote of 74 to 36
- $B$ by a vote of 66 to 44
- $C$ by a vote of 72 to 38
- $D$ by a vote of 56 to 54

So there is a Condorcet winner, namely $E$.

In summary, our political party has employed five different common voting procedures and has come up with five different winning candidates. We see from this illustration that those with the power to select the voting method may well determine the outcome, simply by choosing the voting method.

# Multiple Choice Questions 13

**1.** Consider an election with voting profile

| 4 | 3 | 3 | 3 |
|---|---|---|---|
| B | C | D | C |
| D | A | A | B |
| C | D | B | A |
| A | B | C | D |

Two candidates are to be elected. Who will win under the generalized Hare system?

A.  *B* and *C*              B.  *C* and *D*              C.  *B* and *D*

D.  There is a tie for second; no winners are chosen

**2.** Two candidates are to be elected using the preference profile

| 6 | 5 | 5 | 2 |
|---|---|---|---|
| A | B | C | D |
| D | D | B | B |
| C | C | A | A |
| B | A | D | C |

Who will be chosen under the generalized Hare voting system?

A.  *A* and *B*      B.  *B* and *D*      C.  *A* and *D*      D.  *A*, *B* and *D*

**3.** Eight board members vote by approval voting on four candidates, A, B, C, and D, for new positions on their board as indicated in the following table. An "X" indicates an approval vote.

| Voter | 1 | 2 | 3 | 4 | 5 | 6 | 7 | 8 |
|-------|---|---|---|---|---|---|---|---|
| A | × | × | × | × | × |   | × | × |
| B |   | × |   | × | × | × |   | × |
| C | × |   | × | × |   |   |   | × |
| D | × |   | × | × |   | × | × | × |

Which candidate or candidates will be elected if 70% approval is required and at most two are elected?

A.  Only *A* is elected              B.  Only *D* is elected

C.  Both *A* and *D* are elected      D.  No candidate is elected

**4.** Eight board members vote by approval voting on four candidates, *A*, *B*, *C*, and *D*, for new positions on their board as indicated in the following table. An "X" indicates an approval vote.

| Voter | 1 | 2 | 3 | 4 | 5 | 6 | 7 | 8 |
|-------|---|---|---|---|---|---|---|---|
| A | × | × |   | × |   |   | × | × |
| B |   | × |   | × | × | × |   | × |
| C | × |   | × | × |   |   |   | × |
| D | × |   |   | × |   | × | × | × |

Which candidate or candidates will be elected if 60% approval is required and at most two are elected?

A.   Both *A* and *B* are elected      B.   Both *A* and *D* are elected

C.   Both *B* and *D* are elected      D.   No decision is possible

**5.** The 22 members of the University Senate must choose one of the three alternative schedules, *A*, *B* and *C*. Their preferences are shown. They vote using the Hare system. In this case schedule *A* is chosen. Could the voters who most prefer *C* vote insincerely in some way to change the outcome to one that they would prefer?

| 8 | 4 | 10 |
|---|---|----|
| A | B | C |
| B | A | B |
| C | C | A |

A.   Yes, switch *A* and *B*      B.   Yes, switch *A* and *C*

C.   Yes, switch *B* and *C*      D.   No, *A* still wins

**6.** Thirty-three voters are choosing from among three alternatives, *A*, *B*, and *C*, using a Borda count. Their preferences are shown. If all voters follow their preferences, *C* will win. How many of the seven voters whose favorite is *A* would have to change their preference lists in order to make *B* the winner?

| 7 | 13 | 10 | 3 |
|---|----|----|---|
| A | B | C | C |
| C | C | A | B |
| B | A | B | A |

A.   At least 2      B.   At least 4      C.   All 7

D.   No matter how many change, *C* will still win

## Exercises 13

*In Exercises 1–4, a preference table is shown for an election where the generalized Hare method will be used. Two candidates are to be elected. In each case, what is the quota? What is the outcome of the election?*

**1.**

| 7 | 8 | 7 | 5 |
|---|---|---|---|
| A | B | C | D |
| B | D | B | C |
| C | C | A | B |
| D | A | D | A |

**2.**

| 3 | 5 | 8 | 8 | 6 |
|---|---|---|---|---|
| A | A | B | C | D |
| B | C | D | B | C |
| C | D | C | A | B |
| D | B | A | D | A |

**3.**

| 4 | 6 | 6 | 6 | 2 | 6 |
|---|---|---|---|---|---|
| A | D | B | E | C | B |
| D | A | D | A | E | A |
| E | E | C | D | D | E |
| B | B | E | C | A | C |
| C | C | A | B | B | D |

.

**4.**

| 6 | 7 | 7 | 7 | 7 | 5 |
|---|---|---|---|---|---|
| A | B | C | D | C | E |
| E | D | B | E | D | B |
| D | A | D | A | E | A |
| B | E | E | C | A | C |
| C | C | A | B | B | D |

.

*In Exercises 5–8, a preference table is shown for an election where the generalized Hare method will be used. Three candidates are to be elected. In each case, what is the quota? What is the outcome of the election?*

**5.**

| 4 | 7 | 9 | 8 | 7 | 5 |
|---|---|---|---|---|---|
| A | A | B | C | D | D |
| B | B | D | D | C | A |
| C | D | C | A | B | B |
| D | C | A | B | A | C |

.

**6.**

| 24 | 16 | 19 | 21 | 10 | 10 |
|----|----|----|----|----|----|
| A  | B  | C  | D  | E  | E  |
| B  | A  | D  | C  | C  | A  |
| E  | D  | E  | E  | D  | B  |
| D  | E  | A  | B  | B  | D  |
| C  | C  | B  | A  | A  | C  |

.

**7.**

| 6 | 6 | 6 | 6 | 5 | 3 |
|---|---|---|---|---|---|
| A | A | C | C | E | E |
| B | D | D | E | C | A |
| E | E | E | D | D | B |
| D | B | A | B | A | C |
| C | C | B | A | B | D |

.

**8.**

| 7 | 7 | 7 | 8 | 8 | 3 | 8 |
|---|---|---|---|---|---|---|
| A | A | B | C | C | D | E |
| D | B | C | D | E | C | A |
| E | D | E | E | D | B | B |
| B | E | A | A | A | E | D |
| C | C | D | B | B | A | C |

.

**9.** Eight administrators $s, t, u, v, w, x, y, z$ are voting to fill one position in senior management from candidates $A, B, C, D, E, F$. They use an approval system; their approval table is

|   | s | t | u | v | w | x | y | z |
|---|---|---|---|---|---|---|---|---|
| A | × | × |   |   |   | × |   |   |
| B | × |   | × | × | × |   |   | × |
| C |   | × |   | × |   |   | × | × |
| D |   | × |   | × | × |   | × |   |
| E |   |   | × |   |   |   | × | × |
| F |   | × | × | × |   |   | × | × |

.

(i)   What is the outcome?

(ii)  What is the outcome if at least 80% approval rating is required?

(iii) What is the outcome if at least 60% approval rating is required?

(iv)  The administrators are told that they can appoint more than one person, but at least 60% approval rating is required. What is the outcome?

**10.** Here is the approval table for selection of at most three candidates.

|   | $p$ | $q$ | $r$ | $s$ | $t$ | $u$ | $v$ | $w$ | $x$ | $y$ | $z$ |
|---|-----|-----|-----|-----|-----|-----|-----|-----|-----|-----|-----|
| $A$ | × | × |   | × | × | × |   | × |   |   |   |
| $B$ | × | × | × |   | × | × | × | × |   |   | × |
| $C$ |   | × |   | × |   | × |   |   | × | × | × |
| $D$ |   | × |   | × | × | × |   | × |   | × |   |
| $E$ | × | × | × | × | × | × | × |   | × | × | × |
| $F$ |   |   | × | × |   |   |   | × | × | × |   |
| $G$ |   | × |   |   | × | × |   |   | × | × |   |
| $H$ | × |   | × |   | × | × | × |   |   | × | × |

(i) What is the outcome if there is no minimum requirement?

(ii) How many votes does a candidate need if at least 66% approval is required?

(iii) What is the outcome if at least 66% approval is required?

**11.** Twenty-five electors vote for three candidates, resulting in the preference table

| 8 | 6 | 7 | 4 |
|---|---|---|---|
| $A$ | $B$ | $C$ | $B$ |
| $C$ | $C$ | $A$ | $A$ |
| $B$ | $A$ | $B$ | $C$ |

(i) Who, if anybody, would win under the majority method?

(ii) Who, if anybody, would win under the plurality method?

(iii) What would be the result if a Borda count were used?

(iv) Would any of these results change if the eight voters who preferred $A$ were to exchange their second and third preferences?

**12.** A pairwise sequential election has preference profile

| 2 | 8 | 11 | 7 |
|---|---|----|---|
| $A$ | $C$ | $B$ | $A$ |
| $B$ | $B$ | $A$ | $C$ |
| $C$ | $A$ | $C$ | $B$ |

(i) Who will win under agenda $A,B,C$?

(ii) Who will win under agenda $B,C,A$?

(iii) Who will win under agenda $C,A,B$?

**13.** Thirty committee members with the preference schedules below are to elect their new chairman from among five candidates: $A$, $B$, $C$, $D$, and $E$. If the Borda count is used, candidate $B$ would win.

(i) Would there be any difference in the result if candidate $D$ withdrew from the race before the ranking?

(ii) Would there be any difference in the result if candidate $A$ withdrew from the race before the ranking?

| 8 | 4 | 10 | 5 | 3 |
|---|---|----|---|---|
| A | B | B | D | D |
| D | C | C | B | B |
| B | A | D | A | C |
| C | D | A | E | A |
| E | E | E | C | E |

14. The 22 members of the University Senate must choose one of three alternative schedules, $A$, $B$ and $C$. Their preferences are shown below. They vote using the Hare system. In this case schedule $A$ is chosen. Could the voters who most prefer $C$ vote insincerely in some way to change the outcome to one that they would prefer?

| 8 | 4 | 10 |
|---|---|----|
| A | B | C |
| B | A | B |
| C | C | A |

15. Thirty-three voters are choosing from among three alternatives, $A$, $B$ and $C$, using a Borda count. Their preferences are shown below. Prove that $C$ wins if all voters follow their preferences. How many of the seven voters whose favorite is $A$ would have to change their votes in order to make $B$ the winner (an outcome they would prefer to $C$)?

| 7 | 13 | 10 | 3 |
|---|----|----|---|
| A | B | C | C |
| B | C | A | B |
| C | A | B | A |

16. A committee of 11 members needs to elect one representative from four candidates. They plan to use a Borda count. The preferences are

| 4 | 2 | 2 | 3 |
|---|---|---|---|
| A | A | B | B |
| C | B | C | D |
| D | C | A | C |
| B | D | D | A |

(i) Who would win the election if all electors vote sincerely?

(ii) Can the voters in the third column vote insincerely so as to change the result in their favor? If so, how?

(iii) Can the voters in the fourth column vote insincerely so as to change the result in their favor? If so, how?

17. Suppose an election with the following preference profile is decided by the Hare system.

| 3 | 6 | 7 | 8 | 6 |
|---|---|---|---|---|
| A | A | B | C | D |
| B | C | D | D | C |
| C | D | A | A | B |
| D | B | C | B | A |

(i) Who will win the election?

(ii) Show that the seven supporters of $B$ can achieve a preferred result if they exchange $B$ and $D$ on their ballots (that is, they vote as if their preference was $D,B,A,C$).

18. Consider a runoff election with preferences

| 4 | 4 | 2 | 3 | 4 |
|---|---|---|---|---|
| A | A | B | B | C |
| B | C | A | C | B |
| C | B | C | A | A |

(i) Who wins the election?

(ii) Show that two supporters of $A$ can change the result in favor of their candidate by changing their preferences from $(A,C,B)$ to $(C,A,B)$.

19. A club with 46 members wishes to elect a president. The post is currently held by $Y$. The 46 members have preference profiles

| 16 | 12 | 10 | 8 |
|----|----|----|---|
| X | Y | Z | Y |
| Z | Z | X | X |
| Y | X | Y | Z |

(i) The following electoral system is used: a plurality vote is held between the candidates other than the President; there is then a majority election between the winner of that election and the current President. Show that $X$ will win the election.

(ii) The eight voters whose votes form the right-hand column of the table decide to change their votes in an attempt to make $Y$ the winner. Can they do this?

20. Three committee members must vote to fill one position. There are four candidates, and the preferences are

| A | B | C |
|---|---|---|
| B | C | D |
| C | D | A |
| D | A | B |

They decide to use pairwise sequential voting. For each candidate, find an agenda such that the selected candidate will win.

# Chapter 14
# The Mathematics of Finance

We shall now look at the mathematics of finance: interest, investments, and loans. In our brief overview we can only touch on a few elementary topics. Even for this, a calculator will be essential. In all our financial calculations, we use the idea of percentages. Recall that "percent" means "out of 100." Therefore $R\%$ is just another way of saying $\frac{R}{100}$.

In this chapter we shall introduce simple and compound interest.

## 14.1  Simple Interest

The idea of simple interest is well-known. Suppose you put $100 in a bank account, and at the end of 1 year the bank pays you back your $100 plus $6. This is called *interest* on your investment, and the rate is 6%.

This interest rate is always stated in terms of the annual interest. Suppose you put your $100 in the bank for 3 months, and receive $1.50 in interest. This is 1.5% of your original investment, but the rate is still quoted as 6%, the equivalent rate if the money had been kept for a year. To avoid confusion, it would sometimes be called "6% per annum" ("annum is Latin for "year").

Sometimes you reinvest your money; in the first example, you would have $106 in your account at the beginning of year 2. If this process is carried out automatically by the bank, it is called *compounding*. We shall discuss compounding, and compound interest, in the next section. *Simple interest* is just the case where compounding does not occur. The obvious model is where you take out the interest and spend it, as for example when somebody retires and lives on the interest from their savings. If your investment pays simple interest, and you reinvest the interest rather than spending it, this is essentially the same as compounding.

W.D. Wallis, *Mathematics in the Real World*, DOI 10.1007/978-1-4614-8529-2_14,

The arithmetic of borrowing money (loans, mortgages) is similar to that for investing. Typically, you do not wait until the end of a loan period to pay back a loan. The usual practice is to pay equal amounts each month (or each week or . . . ). For this reason, most loans involve compound interest. However, some loans use simple interest. We shall give examples at the end of this section.

The original amount you borrow is called the *principal*, or *present value* of an investment or loan. Suppose you draw simple interest on a principal $P for $n$ years at $R$ interest. The total amount you would receive is $A, where

$$A = P\left(1 + n\frac{R}{100}\right) = P(1 + nr).$$

(where $r$ is the fraction $r = R/100$; sometimes $r$ is more convenient to use). The total interest is $I, where

$$I = Pn\frac{R}{100} = Pnr.$$

(We shall refer to these two equations as the "first simple interest formula" and the "second simple interest formula.")

When dealing in periods shorter than a year, it is common to calculate as though the year consisted of 12 months, each of 30 days. This 360-day "year" is called a "standard" year. The regular 365-day year is called an "exact" year. Simple interest may be calculated for part of a year—$n$ need not be an integer—and in that case, standard years are normally used.

In the case of an investment, simple interest is usually paid at the end of each year, but sometimes at the end of the loan period; for a loan, both interest and principal are typically paid at the end of the whole loan period. (Compound interest is normally used in cases where parts are paid throughout the loan period.)

**Sample Problem 14.1**   *You borrow $1,600 at 12% simple interest for 4 months. How much must you pay at the end of the period? How much would you pay if the loan were for 2 years?*
**Solution.** We use the formula $A = P(1 + nR/100)$ with $P = 1,600$ and $R = 12$. In the first case, $n$ is $\frac{1}{3}$ (4 months is one-third of a year), so

$$A = 1,600 \times (1 + .12/3) = 1,600 \times (1.04) = 1,664$$

and you repay $1,664. In the second case, $n$ is 2, so

$$A = 1,600 \times (1 + .12 \times 2) = 1,600 \times (1.24) = 1,984$$

and you repay $1,984.
**Your Turn.** You borrow $1,200 at 10% simple interest for 3 months. How much must you pay at the end of the period? How much would you pay if the loan were for 3 years?

## Using the "Simple Interest" Formulas

The first simple interest formula can be rewritten to make $P$ the subject: given the final amount, interest rate and term, you can calculate the principal:

$$P = A\left(1 + n\frac{R}{100}\right)^{-1}.$$

Similarly you can calculate the interest rate from the other data.

**Sample Problem 14.2**  *You borrow $1,800 at simple interest for 6 months. At the end of the period you owe $1,863, including the principal. What was the interest rate?*

**Solution.** Again we use $A = P(1 + nR/100)$. We know $A = 1,863$, $P = 1,800$ and $n = \frac{1}{2}$. So

$$1,863 = 1,800 \times \left(1 + \frac{R}{200}\right) = 1,800 + 9 \times R,$$

and therefore $9 \times R = 63, R = 7$. So the interest rate was 7%.

**Your Turn.** You borrow $1,400 at simple interest for 4 months. At the end of the period you owe $1,428, including the principal. What was the interest rate?

**Sample Problem 14.3**  *You need to borrow some money for 3 months. Your lender offers a rate of 12%. At the end of the period you repay $824. How much was the principal?*

**Solution.** Using the formula with $R = 12$ and $n = \frac{1}{4}$, we get

$$824 = P(1 + 12/400) = P \times 1.03, \ P = \frac{824}{1.03} = 800.$$

The principal was $800.

**Sample Problem 14.4**  *You borrow $1,400 at simple interest for 3 years. At the end of that period, your interest is $210. What was the interest rate?*

**Solution.** In this case $P = 1,400, I = 210$ and $n = 3$. So, using the second simple interest formula,

$$210 = 1,400 \times 3 \times \frac{R}{100} = 42R; R = \frac{210}{42} = 5.$$

The rate was 5%.

**Your Turn.** You borrow $2,200 at simple interest for 2 years. At the end of that period, you owe a total of $2,233. What was the interest rate?

## 14.2   Compound Interest

When dealing with compound interest, students will find it very useful if their calculator enables them to calculate powers of numbers: for example, if it has a key (possibly one marked $x^y$) that enables the user to input two numbers and automatically calculate the result of raising the first number to the power of the second number. (Only positive whole number powers will occur).

Say you have \$100 and every year you double your money.

One year from now you have \$200.

2 years from now you have \$400.

3 years from now you have \$800.

So in 1 year you gain 100%; in 3 years you gain 700%—much more than $3 \times 100\%$. This process is called *compounding*. It also happens for interest less than 100%.

**Sample Problem 14.5**  *Suppose you put* \$1,000 *in the bank for 5 years at* 10% *interest paid annually. If you take your interest out of the bank at the end of each year, how much do you have at the end of 5 years? If you allow it to compound, how much do you have at the end of 5 years?*

**Solution.** If you take your interest out of the bank at the end of each year, you get \$100 each year. After 5 years you have a total of \$1500, a profit of \$500.

If you put your interest back in the bank at 10%:

- After year 1 you get \$100, so you have a total of \$1100 in the bank.
- After year 2 you get \$110 (10% of \$1100), so you have a total of \$1210 in the bank.
- After year 3 you get \$121, for a total of \$1331.
- After year 4 get \$133.10, for a total of \$1464.10.
- After year 5 get \$146.41, for a total of \$1610.51.

So after 5 years you have \$1610.51, a profit of \$610.

Let's look at this in general. Say you put \$P in the bank at $R\%$ for $N$ years, and reinvest all the interest. You end up with

$$P\left(1 + \frac{R}{100}\right)^N.$$

This process is called *geometric growth*.

On the other hand, simple interest is the same as "we'll take the interest out each year." After $N$ years at $R\%$ you would finish with

$$P\left(1 + \frac{RN}{100}\right).$$

This is called *arithmetic growth*.

**Sample Problem 14.6** *Suppose you invest $1,200 at 10% interest for 3 years with interest paid each year. How much interest is earned in total, if you take the interest out each year? How much if you reinvest the interest each year?*

**Solution.** We use the two formulas. For arithmetic growth, you end with

$$1,200 \left(1 + \frac{10 \times 3}{100}\right) = 1,200 \times 1.3 = 1,560$$

and the interest is $(1,560 - 1,200) = \$360$. Under geometric growth, the amount received after 3 years is

$$1,200 \left(1 + \frac{10}{100}\right)^3 = 1,200 \times 1.1^3 = 1,200 \times 1.331 = 1,597.2$$

and the interest is $397.20.

**Your Turn.** What are the results for the above problem if your period of investment is 4 years?

## 14.3 Interest Periods

Very often a bank pays interest more frequently than once a year. Say you invest at $R\%$ per annum for 1 year but interest is paid four times per year. The bank pays $\frac{R}{4}\%$ every 3 months; this is called the *interest period*, or *term*. (Unfortunately, *term* is also used to denote the life of the loan.) If you invest $A$, your capital after a year is

$$\$P \left(1 + \frac{R}{400}\right)^4,$$

just as if the interest rate was divided by 4 and the number of years was multiplied by 4.

(Notice that we used the simplifying assumption that 3 months is a quarter of a year. In fact some quarters can be 92 days, others 91 or 90, but banks seldom take this into account. Similarly, when interest is calculated every month, it is usual to assume that each month is one-twelfth of a year.)

**Sample Problem 14.7** *Say your bank pays $R = 8\%$ annual interest, and interest is paid four times per year. What interest rate, compounded annually, would give you the same return?*

**Solution.** $\left(1 + \frac{R}{400}\right)^4 = (1.02)^4 = 1.0824\ldots$ so the effect is the same as compounding annually with $8.24\ldots\%$ interest rate.

Say interest is added $t$ times per year. The result after 1 year is the same as if the interest rate were $(R/t)\%$ and the investment had been held for $t$ years: after 1 year,

$$P\left(1 + \frac{R}{t \times 100}\right)^t,$$

so after $N$ years,

$$P\left(1 + \frac{R}{t \times 100}\right)^{tN}.$$

The calculations are exactly the same when you borrow money as they are when you invest money.

**Sample Problem 14.8** *You borrow $50,000 at 10% annual interest, compounded every 3 months, for 10 years. Assuming you make no payments until the end of the period, how much will you owe (to the nearest dollar)?*
**Solution.** We have $P = 50,000$, $R = 10$, $t = 4$ and $N = 10$. So

$$P\left(1 + \frac{R}{t \times 100}\right)^{tN}$$

becomes

$$50,000 \times \left(1 + \frac{10}{400}\right)^{40} = 50,000 \times (1.025)^{40} = 50,000 \times 2.68506$$

which comes to $134,253.19\ldots$ and you owe $134,253.
**Your Turn.** You borrow $40,000 at 12% annual interest, compounded every 3 months, for 15 years. Assuming you make no payments until the end of the period, how much will you owe?

In Sample Problem 14.7, 8% is called the *nominal rate* of interest. The nominal rate doesn't tell you how often compounding takes place. $8.24\ldots\%$ per annum is the *effective rate*. (These terms are established in the Truth in Savings Act.) Sometimes banks and other lenders will talk about nominal rates per month or per quarter, but we shall always use these terms for the annual rates unless we specify otherwise.

When discussing a loan, the annual nominal rate is also called the *annual percentage rate* or APR. The yearly effective rate is called the *annual percentage yield* (APY) or *effective annual rate* (EAR). To avoid confusion, we shall refer to the APY, both for investments and loans.

Banks and other lenders like to tell you the APR, but what you really want to know is the APY. To calculate the APY, one works out how much would be owed on $100 at the end of a year, if no payments were made. This is not a real-world calculation: credit card companies and mortgage-holders normally require some minimum payment, or a penalty is charged. In calculating the APY, act as though all penalties are waived.

**Sample Problem 14.9** *Your credit card company charges an APR of 18%. Payments are required monthly, and interest is charged each month. What is the corresponding APY?*
**Solution.** The amount owing from a $100 loan at the end of 1 year is

$$A = 100\left(1 + \frac{18}{12 \times 100}\right)^{12}$$

$$= 100 \times 1.015^{12}$$

$$= 119.56,$$

so the APY is 19.56%.
**Your Turn.** Suppose your credit card company charges an APR of 12%, under the above conditions. What is the corresponding APY?

We said earlier that it is usual to use the 360-day "standard" year instead of the more accurate 365-day model. To conclude this chapter we show that this does not make a great deal of difference. Suppose you were to borrow money at the rate of 5%, compounded daily. If a 360-day year is used, the APY is

$$(1 + .05/360)^{360} - 1 = (1.00013888\ldots)^{360} - 1 = .0512671\ldots$$

or 5.12671...%, while the 365-day APY is

$$(1 + .05/365)^{365} - 1 = (1.00013699\ldots)^{365} - 1 = .0512675\ldots$$

or 5.12677...%. If you borrow $10,000 and repay your loan after a year, you will owe between $10,512.67 and $10,512.68. The difference is less than one cent. And, in the real world, your debt would probably be rounded to the nearest dollar.

# Multiple Choice Questions 14

1. If you deposit $8,000 at 7% simple interest, what is the balance after 5 years?
   A.  $8,560.00       B.  $10,579.60       C.  $10,800.00       D.  $11,318.22
2. You have $48,600 that you invest at 5% simple interest. How long will it take
   for your balance to reach $85,050?
   A.   14 years       B.   15 years       C.   16 years       D.   17 years
3. John invests $4,000 at 4% simple interest. If he does not withdraw any of the
   principal or interest earned during the next 10 years, how much is the account
   worth at the end of the 10 years?
   A.  $5,920.00       B.  $4,162.50       C.  $5,600.00       D.  $6,070.00
4. If you deposit $8,000 at 7% compounded quarterly, what is the balance after
   5 years?
   A.  $8,560.00       B.  $10,579.60       C.  $10800.00       D.  $11,318.22
5. If you deposit $8,000 at 7% compounded monthly, what is the balance after
   5 years?
   A. $11,341.00       B. $11,318.22       C. $11,220.41       D. $463,571.41
6. Margaret borrowed $1,000 from her parents, agreeing to pay them back when
   she graduated from college in 5 years. If she paid interest compounded quarterly
   at 5%, how much would she owe at the end of the 5 years?
   A.   $1,050        B.   $1,282        C.   $1,503        D.   $1,581
7. What is the APY for 7.5% compounded quarterly, to one decimal place?
   A.   7.3%           B.   7.5%           C.   7.7%           D.   8.1%
8. What is the APR for 7.5% compounded quarterly?
   A.   7.3%           B.   7.5%           C.   7.7%           D.   8.1%
9. Which of these loan rates is most favorable to the lender?
   A.   12% compounded annually       B.   12% compounded quarterly
   C.   12% compounded monthly        D.   12% compounded daily

# Exercises 14

1. You borrow $4,000 at 8% simple interest. What is the total you must pay if the loan is for a period of:
   (i)   1 year;                    (ii)   3 years;                    (iii)   4 years?

2. You borrow $8,000 at 3% simple interest. What is the total interest if the loan is for a period of:
   (i)   two years;                 (ii)   three years;               (iii)   six years?

3. You borrow $2,500 at 5% simple interest. What is the total you must pay if the loan is for a period of:
   (i)   three years;               (ii)   four years;                (iii)   seven years?

4. You borrow $3,000 at 5% simple interest. What is the total interest if the loan is for a period of:
   (i)   two years;                 (ii)   four years;                (iii)   five years?

5. You borrow $2,000 at 3% simple interest. Assume the interest is calculated on a standard (360-day) year. What is the total you must pay if the loan is for a period of:
   (i)   1 month;                   (ii)   4 months;                  (iii)   6 months?

6. You borrow $4,800 at 5% simple interest. Assume the interest is calculated on a standard (360-day) year. What is the interest if the loan is for a period of:
   (i)   1 month;                   (ii)   3 months;                  (iii)   6 months?

7. You borrow $400 at simple interest for 3 years. At the end of the loan, you owe $120 in interest. What was the interest rate?

8. You borrow $600 at simple interest for 2 years. At the end of the loan, you must repay $690 in total. What was the interest rate?

9. You borrow $1,000 at simple interest for 6 months. At the end of the loan, you owe $60 in interest. What was the interest rate?

10. You borrow $5,400 at simple interest for 9 months. At the end of the loan, you must repay a total of $5,562. What was the interest rate?

11. You borrow some money at 8% simple interest. How much did you borrow, if:
    (i)   the interest you owe at the end of 4 years is $1,344;
    (ii)  your total repayment at the end of 8 months is $13,272;
    (iii) the interest you owe at the end of 6 months is $44?

12. You borrow some money at 6% simple interest, calculated on a standard (360-day) year. How much did you borrow, if:
    (i)   your total repayment at the end of 2 years and 4 months is $3,648;
    (ii)  your total repayment at the end of 8 months is $2,912;
    (iii) the interest you owe at the end of 6 months is $198?

13. A company borrows $17,300 at 6.8% simple interest to cover short-term costs, and pays interest of $882.30. How long was the period of the loan?

14. You borrow $1,000 at 12% interest, and repay it after 1 year. What is the total payment if the interest is compounded
    (i)   every 3 months;           (ii)   every 6 months;            (iii) once a year?

15. You borrow $100 at 6% interest for 4 years. What is the total interest if compounding takes place
    (i) every month;        (ii) every 3 months;        (iii) once a year?
16. You invest $3,250 at 2% interest compounded annually. What is its value after 5 years?
17. You borrow $625 at 8% interest compounded quarterly, and repay it after 12 years. How much interest must you pay?
18. You invest $200 for 1 year at interest rate 12%, compounded monthly. What is the value of your investment at the end of the year?
19. You invest $2,000 for 3 years at interest rate 6%, compounded every 6 months. What is the value of your investment at the end of the period?
20. You invest a sum for 12 years at interest rate 12%, compounded quarterly. At the end of the period your investment is worth $10,000. How much did you invest initially (to the nearest dollar)?
21. You wish to deposit a sum at 6% interest, compounded every 6 months, in order to pay $10,000 due in 5 years. How much must you deposit (to the nearest dollar)?
22. You deposit $4,000 at 7% interest, compounded monthly. How many years will it take until your investment exceeds $9,000?
23. You invest your money at 12% compound interest paid quarterly. When will it double in value?
24. What is the APY on a loan:
    (i)  at 5% APR, compounded monthly;
    (ii) at 6% APR, compounded quarterly;
    (iii) at 3% APR, compounded monthly?

# Chapter 15
# Investments: Loans

You do not normally wait until the end of a loan period to pay back a loan. The usual practice is to pay equal amounts each month (or each week or ...). Another situation in which equal deposits are made is the periodic savings account, such as a Christmas club or retirement account, where a fixed amount is deposited into savings each period.

## 15.1 Regular Savings

Consider a periodic savings account. Suppose you deposit $D$ each month. The interest each month is $M\%$; write $m = M/100$. Assume the account is empty to start, and you pay in for $n$ months. (Often $n = 11$ or $12$, because people use these accounts to save for vacations or Christmas shopping.)

The calculations to find the amount at the end of the $n$-th month might start:

| | |
|---|---|
| Principal, start of month 1 | $D$ |
| Interest earned in month 1 | $mD$ |
| Principal, end of month 1 | $(1+m)D$ |
| | |
| Add to principle | $D$ |
| | |
| Principal, start of month 2 | $D + (1+m)D$ |
| | $= (2+m)D$ |
| Interest earned in month 2 | $m(2+m)D$ |
| Principal, end of month 2 | $(2+m)D + m(2+m)D$ |
| | $= (1+m)(2+m)D$ |
| | |
| Add to principle | $D$ |
| | |
| Principal, start of month 3 | $D + (1+m)(2+m)D$ |

and so on.

W.D. Wallis, *Mathematics in the Real World*, DOI 10.1007/978-1-4614-8529-2_15,     215
© Springer Science+Business Media New York 2013

This soon becomes complicated. An easier way is to calculate the effect of putting each new payment in a new bank account. The total in all the accounts at the end of $n$ months will be the required amount.

Payment 1 draws interest for $n$ months, so the amount in that account at the end is $\$D(1+m)^n$; payment 2 draws interest for $n-1$ months, so the amount in that account at the end is $\$D(1+m)^{n-1}$. The total of accounts 1 and 2 is

$$\$D(1+m)^n + \$D(1+m)^{n-1}$$
$$= \$D[(1+m)^n + (1+m)^{n-1}]$$

If we proceed in this way, the total after $n$ months ($n$ accounts) is

$$\$D\left[(1+m)^n + (1+m)^{n-1} + \ldots + (1+m)\right].$$

If $m = 0$ then the total is simply $\$nD$. We assume $m \neq 0$ and evaluate

$$(1+m)^n + (1+m)^{n-1} + \ldots + (1+m).$$

Write

$$X = (1+m)^n + a^{n-1} + \ldots + (1+m)^2 + (1+m).$$

Then

$$(1+m)X = (1+m)^{n+1} + (1+m)^n + \ldots + (1+m)^3 + (1+m)^2.$$

Subtracting, $mX = (1+m)^{n+1} - (1+m)$ so $X = \frac{(1+m)^{n+1}}{m}$ (this is where we need to assume $m \neq 0$).

So the amount in the account after $n$ months is

$$\frac{\$D \times [(1+m)^{n+1} - (1+m)]}{(1+m) - 1}$$
$$= \frac{\$D \times [(1+m)^{n+1} - (1+m)]}{m}$$

We call this amount the *accumulation*.

**Sample Problem 15.1** *At the beginning of each month you put $100 into an account that pays 6% annual interest. How much have you accumulated at the end of the year?*

**Solution.** 6% annual interest is .5% per month. So $m = .005, n = 12, D = 100$, and you get

$$\$100(1.005^{13} - 1.005)/.005$$
$$= \$100(1.066986 - 1.005) \times 200$$

$$= \$20000(0.061986)$$

$$= \$1239.72$$

**Your Turn.** In the above example, suppose you started saving in April, so that you only made nine payments. How much will you accumulate?

Some investment funds are set up so that you make your payment at the *end* of the payment period, rather than the beginning. In these cases it is usual to add the last payment to the accumulation, even though it accrues no interest. In that case the accumulation is

$$\frac{\$D \times [(1+m)^n - 1]}{m}.$$

For example, in a Christmas club, you might make your first payment on January 31st and withdraw the money late in November. There are ten payments. If the annual interest is again 6%, your accumulation is

$$\frac{\$D \times [(1.005)^{10} - 1]}{.005}.$$

## 15.2   Compound Interest Loans and Payments

When you borrow money at compound interest and make regular repayments, interest is normally calculated on the amount you owe. Usually the interest is calculated at the time your payment is due; when you buy a house on a mortgage and make monthly payments, the interest is compounded monthly. There is a penalty for late payments, in addition to the interest on the missed payment. Usually there is no reward for paying early in the month. We shall refer to this arrangement as a *standard compound interest loan*, or simply a *compound interest loan*.

To clarify this, suppose your house payment of $2,000 is due on the first of the month, and on June 1st this year your total indebtedness is $100,000. For simplicity, say your annual interest rate is 12%, so the interest for 1 month is 1%. On July 1st, interest of $1,000 (1% of $100,000) is added to your debt. Then payments are credited. Assuming you made the standard $2,000 payment, this is subtracted from your debt, which becomes ($100,000+$1,000−$2,000) = $99,000. When August 1st comes around, the new interest will be $990 (1% of $99,000). It does not matter when you made the payment, provided it is on or before July 1st; even if you paid on June 2nd, it is applied on July 1st. If you made a payment greater than $2,000 during June, the total would be subtracted from your debt on July 1st.

If you miss a payment, or pay less than $2,000, some penalty is exacted. The arrangements differ from loan to loan. Some lenders charge a higher interest rate on the amount in arrears; some charge a fee; many do both.

The amount of payment required to pay off a loan can be calculated from the data about the loan.

The calculation proceeds as follows. Say you borrow $P$ at $M\%$ monthly interest (compounded monthly), and pay it back at the end of $Y$ years. The arithmetic is the same as if you put $D$ into a savings account each month at $M\%$ interest compounded monthly, and at the end of $Y$ years you have exactly enough money to pay off your loan. This amount is $P(1+m)^n$, where $n = 12Y$ and $m = M/100$. Then the required monthly payment is $D$.

The only difference between this and the example of accumulated savings is that loan repayments usually start at the *end* of the first month, so for $Y$ years the total number of months for which your money accumulates is $12Y - 1$, not $12Y$, and you gain nothing during the first month. If we continue to write $n$ for the number of months in the life of the loan, we use $n - 1$ in place of $n$ in the accumulation formula. This actually simplifies the formula: we want to sum

$$D\left[(1+m)^{n-1} + (1+m)^{n-2} + \ldots + 1\right].$$

the required payment $D$ for a loan is calculated from

$$P(1+m)^n = \frac{D \times \left[(1+m)^n - 1\right]}{m}$$

**Sample Problem 15.2**  *You take out a compound interest loan of $100,000 at 6% annual interest to pay off your house. The period is 30 years. What payment is required each month?*

**Solution.** Suppose the monthly payment is $D$. The interest rate is $.5\% = .005$ per month. There are 360 months in 30 years. So $P(1+m)^n$ is

$$\$100,000 \times (1.005)^{360}$$

or $602,257.52$. In this example

$$\frac{D \times \left[(1+m)^n - 1\right]}{m} = \frac{\$D \times \left[(1.005)^{360} - 1\right]}{.005}$$

$$= \$D \times 5.0225752 \times 200 = \$D \times 1004.515$$

so $D \times 1004.515 = 602,257.52$ and $D = 599.55$. You would pay $599.55 per month.

**Your Turn.** To buy your car, you borrow $12,000 over 5 years at 8% interest, compounded monthly. To pay it off you pay $D$ per month. What is $D$?

It is interesting to observe the differences that follow from small changes in a long-term loan. Suppose we change the annual rate of interest in the preceding example from 6% to 8%, leaving the principal and period unchanged. The interest rate is $.00666\ldots$ or $1/150$ per month. So the accumulation after 30 years at 8% is

$$\$100,000 \times (151/150)^{360}$$

or $1,093,572.96. In the "regular savings" model, depositing $D per month, your accumulated savings would be

$$\frac{\$D \times [(151/150)^{360} - 1)]}{1/150} = \$D \times 9.9357296 \times 150 = \$D \times 1490.36$$

So your monthly payment is $D = 733.76$.

On the other hand, changing the period of the loan makes less difference than you might think. Reducing the period by 20%, to 24 years, adds less than 10% to your monthly payment:

**Sample Problem 15.3** *You take out a compound interest loan of $100,000 at 6% annual interest to pay off your house. The period is 24 years. What payment is required each month?*

**Solution.** As in Sample Problem 15.2, the interest rate is .5% $= .005$ per month. There are 288 monthly payments. So in this case $P(1+m)^n$ is

$$\$100,000 \times (1.005)^{288}$$

or $420,557.89, and

$$\frac{D \times [(1+m)^n - 1]}{m} = \frac{\$D \times [(1.005)^{288} - 1)]}{.005}$$

$$= \$D \times 3.2055789 \times 200 = \$D \times 641.116.$$

So $D \times 641.116 = 420,557.89$. You would pay $655.98 per month.

**Your Turn.** Repeat the preceding calculation when the interest rate is 8%.

Here is another way of interpreting the above calculation: suppose you contracted to buy your $100,000 house at 6% annual interest over 30 years. Your bank requires a monthly payment of $600 (actually $599.55 per month, but banks often round up slightly, and reduce the last payment a little). If you decided to pay an extra $56 each month, you would finish paying for your house *6 years* ahead of schedule.

We have calculated $D$ from $A$ (answering the question "What payment must I make?"). We can also calculate $A$ from $D$ ("Given the maximum payment I can make, how much can I afford?")

**Sample Problem 15.4** *You want to buy a car. You can get an 8% loan over 5 years. You can pay $200 per month. How much can you afford to pay for the car?*

**Solution.** $m = 8/12\% = 1/150, D = 200, n = 60$. So

$$mP(1+m)^n = D \times [(1+m)^n - 1]$$

becomes

$$P(151/150)^{60}/150 = 200[(151/150)^{60} - 1],$$

So $P \times 1.49646 = 150 \times 200 \times 0.49646, P = 9,952.69$, and you can afford about $9,950.

**Your Turn.** In the above example, suppose you negotiated a loan at 7.5%. How much could you then afford?

## 15.3  Some Other Loans

We shall look at two forms of loan that are calculated using simple interest, but involve higher interest payments than you might expect.

The first example is the *add-on loan*. In an add-on loan you add the whole amount of simple interest to the principal at the beginning of the loan period. If the principal is $P$, the interest is $R\%$, and the period is $n$ years, then the total to be paid back is $P(1 + \frac{nR}{100})$.

Very often these loans are for short periods, and in those cases the interest is high.

**Sample Problem 15.5**  *What is your monthly payment on an add-on loan if you borrow $12,000 over 5 years at 8% per year?*

**Solution.** Simple interest is $960 per year, so the total (simple) interest for 5 years is $4,800. Therefore the total to be paid is $16,800. There are 60 monthly payments, so your monthly payment will be $16,800/60, which is $280.

**Your Turn.** What is your monthly payment on an add-on loan if you borrow $500 over 3 months at 36% per year?

The second example is a *discounted loan*. In a discounted loan you subtract the interest from the amount borrowed. Suppose your loan says you will borrow $P$ at an interest rate of $R\%$, and the period is $n$ years. Instead of $P$ you receive $P(1 - Rn/100)$. For example, if your loan has principal $12,000 over 5 years at 8%, you only receive

$$\$12,000 \times (1 - .4) = \$7,200.$$

At the end of the period you repay the original principal, $12,000 in the example.

These loans are sometimes used by auto sales companies, for lease agreements with an option to buy.

**Sample Problem 15.6**  *You need to pay $12,000. What will be your payments for a 5-year discounted loan at 8% per year?*

**Solution.** If you need $12,000 then your principal will be $A$ where

$$A \times (1 - .4) = 12,000$$

so $A = \$(12,000/0.6) = 20,000$. Your monthly payments total $20,000 over 60 months, so they equal $333.33 per month, to the nearest cent. (In actual fact, you would probably pay $333.34 per month, with a last payment of $332.94, or maybe $334 per month, with the last payment adjusted down.)

Observe the difference between the payments in the two Sample Problems. This is not an isolated example. An add-on loan is always better than a discounted loan at the same (non-zero) interest rate.

To see this, suppose you need $100. If you borrow $100 for $n$ years at $R\%$ interest, using an add-on loan, you eventually pay $100(1 + \frac{nR}{100}) = \$(100+nR)$. In order to obtain $100 using a discounted loan at $R\%$, your "principal" is $P$, where $P(1 - nR/100) = 100$.

Suppose the discounted loan were as good a deal as the add-on. Then $P \leq 100 + nR$. Then

$$100 = P(1 - nR/100) \leq (100+nR)(1 - nR/100) = (100+nR)(100 - nR)/100$$

from which

$$10,000 \leq (100+nR)(100 - nR) = 10,000 - n^2R^2.$$

This would mean $n^2R^2 \leq 0$. This is never true.

In any case, for a discounted loan or an add-on loan to be worthwhile, the interest rate must be low. They are better for short-term loans.

One can compare compound interest loans with these other sorts of loans. For example, an add-on loan of $1,000 at 5% interest for a 4-year period requires monthly payments of $25. If you took out a loan for 4 years at 6% compound interest, and found your monthly payment to be $25, your principal was $P$, where

$$.005 \times P \times 1.005^{48} = 25 \times [1.005^{48} - 1].$$

Then

$$P = \frac{25 \times [1.005^{48} - 1]}{.005 \times 1.005^{48}}$$

$$= \frac{6.76223}{.0063525}$$

$$= 1,064.50.$$

So you could have borrowed $64.50 more under compound interest at 6%, for the same repayment. In other words, the APY for the 4-year 5% add-on loan is greater than 6%.

## 15.4   Equity

Suppose you have finished 3 years' payment on a 5-year loan of $9,952 at 8% annual interest, for a car. As we saw in Sample Problem 15.4, your payments were $200 per month.

Think about the remaining 24 months. Your situation is as though you had just taken a loan of $A$ at 8% per annum, where

$$A(151/150)^{24}/150 = 200[(151/150)^{24} - 1],$$

so $1.17288A = 150 \times 200 \times 0.17288, A = 4,421.94$.

We say your *equity in the loan* is $9,952 - $4,421.94 , about $5,530.

You might think that, after making payments for three-fifths of the payment period, you would own 60% of your car. However your equity is a little less than that amount: around 55.5%.

The difference is *much* greater on longer-term loans. For example, suppose you take out a 30-year house loan for $100,000 at 8% per annum, with equal monthly payments of $733.76 (as we calculated above). After three-quarters of the term— 270 of the 360 payments have been made—it is as if you had just taken a loan at 8% per annum with principal $A$, where

$$A(151/150)^{90}/150 = 733.76[(151/150)^{90} - 1],$$

that is

$$A \times .012123295 = 600.578$$

so $A = 49,539.20$ and your equity is $50,460.80.

After three-quarters of your payments, you own about half of your house.

# Multiple Choice Questions 15

1. You deposit $170 from your paycheck at the end of each month into a savings account earning 9% interest compounded monthly. How much is the account worth at the end of the 10 years (to the nearest dollar)?
   A.  $4,160          B.  $4,992          C.  $32,900          D.  $42,300

2. A woman makes a $20 deposit each month into a savings account that earns 8% interest compounded quarterly. If she does this for 5 years, how much will the account be worth? (Answer to the nearest dollar.)
   A.  $1,487          B.  $1,654          C.  $1,904          D.  $2,112

3. An annuity has monthly deposits of $700, with 4% annual interest compounded monthly. What is its value in 5 years? (Answer to the nearest dollar.)
   A.  $27,442         B.  $26,410         C.  $46,584         D.  $61,414

4. How much would you have to invest each month in an annuity earning 5% per annum, compounded monthly, in order to earn $200,000 at the end of 18 years? Give your answer to the nearest dollar, rounded up.
   A.  $573            B.  $1,442          C.  $1,593          D.  $1,611

5. You take out a compound interest loan for $12,000 to purchase a car. The interest rate is 8.3% compounded monthly and you have 8 years to repay the loan. What are your monthly payments?
   A.  $171.48         B.  $186.49         C.  $305.21         D.  $385.07

6. A student loan of $12,000 has an interest rate of 7% compounded monthly. Payments are monthly and the borrower has 10 years to repay. What are the monthly payments?
   A.  $95.46          B.  $139.33         C.  $169.53         D.  $290.15

7. You borrow $1,200 in a 9% add-on loan that is to be repaid in equal monthly installments over 2 years. How much will you pay each month?
   A.  $60.00          B.  $41.66          C.  $59.00          D.  $50.00

8. A 7% add-on loan of $18,000 is to be repaid in monthly installments over 6 years. How much is each monthly payment?
   A.  $205            B.  $217            C.  $250            D.  $355

9. You need $24,000, using a discounted loan at 4% interest, to be paid back in equal installments over 10 years. How much will you need to borrow initially?
   A.  $25,000.        B.  $40,000.        C.  $43,000.        D.  $44,500

10. You need $20,000. You take a discounted loan at 4% interest for 5 years. How much do you need to borrow?
    A.  $16,000        B.  $20,000         C.  $24,000         D.  $25,000

# Exercises 15

1. You invest $200 every quarter for 20 years in an annuity that pays 5% interest compounded quarterly. What is the final value of the annuity?

2. You invest $4,000 every year for 5 years in an annuity that pays 10.5% interest compounded annually. What is the final value of the annuity?

3. You invest $2,000 every 6 months for 10 years in an annuity that pays 8% interest compounded twice yearly. What is the final value of the annuity?

4. You need to have $100,000 in 10 years, so you set aside a fixed sum every 3 months in a savings account. How much should you set aside each quarter if the interest is:

    (i)  6%;                    (ii)  8%;                    (iii)  10%?

5. You take out a compound interest loan of $200,000 at 6% annual interest to pay off your house. The period is 30 years. What payment is required each month?

6. Credit card interest is 18% interest compounded monthly. How much must be paid each month to eliminate a debt of $1,000 in 1 year?

7. $200 is invested per month in a fund that pays 9% interest compounded monthly. The first payment is made at the end of the first month. What is the value of the annuity after:

    (i)  1 year;        (ii)  3 years;        (iii)  5 years;        (iv)  8 years?

8. A house mortgage is set at 9%, compounded monthly. If the house costs $200,000, what is the monthly payment if the term of the mortgage is

    (i)   15 years;                        (ii)   30 years?

9. You take out a compound interest loan of $120,000 at 6.5% annual interest to pay off your house. The period is 30 years. What payment is required each month?

10. You want to buy a car for $8,000. The dealer offers you a 5% add-on loan for 4 years, with monthly payments. You can borrow $8,000 for 4 years from your credit union at 7.5% interest. What would be your monthly payment in each case? Which is the better deal?

11. You take out a compound interest loan of $100,000 at 6% annual interest to pay off your house. The period is 30 years. (We saw earlier that your monthly payment is $599.55.) What is your equity after:

    (i)   15 years;                    (ii)   20 years;                    (iii)   22.5 years?

12. What is your monthly payment on an add-on loan if you borrow $3,000 over 3 years at 12% per year?

13. What is your monthly payment on an add-on loan if you borrow $1,200 over 6 months at 36% per year?

14. You take out a compound interest loan of $100,000 at 6% annual interest to pay off your house. The period is 24 years. As we saw, your monthly payment is $655.98. What is your equity after:

    (i)   15 years;                        (ii)   20 years?

15. You need to pay $24,000. How much must you borrow for a 5-year discounted loan at 6%? What will be your monthly payments?

16. You need to pay $16,000. How much must you borrow for a 3-year discounted loan at 9%? What will be your monthly payments?

# Chapter 16
# Growth and Decay

In this short chapter we examine another aspect of compounding, in which either the compounding period is very short or the number of periods is very great. Again you may wish to have a calculator that enables you to raise numbers to integer powers while studying this section. It is also useful to have the constant $e$ available on your calculator.

## 16.1 Continuous Compounding

In some cases, compounding takes place after a very short interval. For example, in some bank accounts, interest is calculated every day. Over 1 year, there are 365 compounding periods. (We saw earlier that the interest rate differs very little between the final interest if the standard year is used instead of the exact year.) In other cases, even though the period is longer, there are still a large number of compounding periods. For example, suppose a company invests some of its funds and loses track of the investment for 100 years. It is found that there is little difference whether compounding took place monthly or quarterly.

For our calculations it will be convenient to work in terms of the fraction $r = R/100$ rather than the percentage $R$. If your bank applies interest to your account $n$ times per year, the interest is $R\%$ and your initial investment ("capital") was $\$A$, then your ending capital after $N$ years is

$$\$A(1 + r/n)^{nN}.$$

Consider the example of 100% interest, the case $r = 1$. The following table shows the values of $(1 + 1/n)^n$ and $(1 + 1/n)^n - 1$ for several values of $n$. The right-hand number is the APY corresponding to an APR of 100% with compounding $n$ times annually.

W.D. Wallis, *Mathematics in the Real World*, DOI 10.1007/978-1-4614-8529-2_16,

| $n$ | $(1+1/n)^n$ | |
|---|---|---|
| 1 | 2.0000000 | 1.0000000 |
| 2 | 2.2500000 | 1.2500000 |
| 5 | 2.4883200 | 1.4883200 |
| 10 | 2.5937424 | 1.5937424 |
| 100 | 2.6915880 | 1.6915880 |
| 1000 | 2.7169239 | 1.7169239 |
| 10000 | 2.7181459 | 1.7181459 |
| 100000 | 2.7182682 | 1.7182682 |

Eventually the right-hand number gets very close to a fixed value, usually denoted $e$, about $2.7182818\ldots$

A similar calculation can be made for other values of $r$. For large $n$, $(1+r/n)^n$ gets very close to $e^r$. For example, if $r = .1$, then $e^r = 1.105171\ldots$. Look at the result when we calculate $(1+r/n)^n$ for various values of $n$:

$$
\begin{aligned}
n &= 1 & (1+r/n)^n &= 1.1^1 & &= 1.1 \\
n &= 10 & (1+r/n)^n &= 1.01^{10} & &= 1.104622\ldots \\
n &= 100 & (1+r/n)^n &= 1.001^{100} & &= 1.105116\ldots \\
n &= 365 & (1+r/n)^n &= 1.000274\ldots^{1}00 & &= 1.105156\ldots \\
n &= 1000 & (1+r/n)^n &= 1.0001^{1}000 & &= 1.105165\ldots
\end{aligned}
$$

We define *continuous compounding* to be a process where, if your original capital is $\$P$, then after $N$ years, the ending capital is

$$\$Pe^{rN}.$$

The calculation above, case $n = 365$, shows that continuous compounding is a good approximation to daily interest, and many lenders use it instead of daily interest because it is easy to calculate.

**Sample Problem 16.1** *You borrow $\$50,000$ at 10% annual interest, compounded continuously, for 10 years. Assuming you make no payments until the end of the period, how much will you owe (to the nearest dollar)?*
**Solution.** We have $P = 50,000$, $R = 10$ and $N = 10$. So $r = .1$ and

$$Pe^{rN} = 50,000 \times e = 135,914.09\ldots$$

and you owe $\$135,914$. Compare this with the result of Sample Problem 14.8, where compounding was quarterly and the answer was $\$134,253$. The difference is not great.

**Your Turn.** You borrow $\$40,000$ at 12% annual interest, compounded continuously, for 15 years. Assuming you make no payments until the end of the period, how much will you owe?

The calculations involved in continuous compounding are closely related to *natural logarithms*. The logarithm to base $b$ of a number $x$, written $\log_b(x)$, is that number $n$ such that $b^n = x$. For example, $\log_b(b) = 1, \log_b(b^2) = 2$, etc. If $b^n = x$ and $b^m = y$, then $xy = b^{n+m}$, so logarithms can be used to simplify arithmetic: in order to multiply two numbers, you add their logarithms. The most commonly used logarithms are to base 10; if you know that $\log_{10}(2) = 0.3010\ldots$, then $20 = 10 \times 2$ so $\log_{10}(20) = \log_{10}(10) + \log_{10}(2) = 1.3010\ldots, \log_{10}(200) = 2.3010\ldots$, and so on. If we just write $\log(x)$ it is assumed that the base is 10. A natural logarithm is a logarithm to base $e$. (The natural logarithm function, written *ln*, is available on most scientific calculators.)

One advantage of continuous compounding is that one can easily calculate the interest when principal and interest rate are known. For example, suppose you invest \$100 for 2 years and receive \$21 interest. Your money grew from \$100 to \$121 with continuous compounding, so the rate $r$ satisfied $100e^{2r} = 121$, so $e^{2r} = 1.21$ and $e^r = \sqrt{1.21} = 1.1$. From this, $r$ will equal $\log_e(1.1)$, which equals $.0953\ldots$, so the interest rate was 9.53%.

The natural logarithm function, written *ln*, is available on most scientific calculators. However, you will seldom need to look it up. You don't even need to calculate the value of $r$ in order to find the interest; all you need is $e^r$. In the example, after $n$ years, your capital will be $\$100 \times (1.1)^n$ and your interest will be $\$100 \times [(1.1)^n - 1]$. Of course, you can easily calculate the APY—in the example, your money would grow from \$100 to \$110 in the first year, so the APY is 10%. If you don't want to deal with natural logarithms, there are tables to calculate the value of $R$ (and $r$) from the APY.

**Sample Problem 16.2**  *You borrow \$50,000 at continuous compounding for 10 years. At the end of the period, you owe \$100,000. What was the approximate APY?*

**Solution.** Suppose the rate is $r$. Then $50,000e^{10r} = 100,000$ so $e^{10r} = 2$. Therefore $e^r$ equals the tenth root of 2, which equals 1.0718 approximately. The amount owed at the end of 1 year—principal plus 1 year's interest—is about 1.0718 times the principal. So the APY is approximately 0.0718, or 7.18%.

**Your Turn.** You borrow \$40,000 for 7 years. At the end of the period, you owe \$80,000. What was the approximate APY?

Observe that the figures of \$50,000 and \$100,000 were not an essential part of the above Sample Problem. The main point was the *ratio*: the debt at the end was double the principal. The answer would be the same for any value of $P$, provided the final amount owed was twice $P$.

If $k$ is any constant, the function $f(x) = k^x$ is called an *exponential function*, and the situation where a quantity changes over $x$ units of time from $A$ to $Ak^x$ is called *exponential growth*; $x$ is the *exponent*. So continuous compounding is an example of exponential growth.

## 16.2   Inflation: The Consumer Price Index

There is a tendency for the purchasing power of money to decrease over time; this is called *inflation*. Sometimes the rate of inflation will change rapidly, but often it stays roughly constant for several years. Inflation follows the same model as continuous compounding.

For example, let's suppose inflation rate is 5% from 2012 to 2017. If something cost $100 in January 2012, then 5 years later, in January 2017, we expect it to cost $100 \times (1.05)^5 = \$127.63$. Of course, individual items do not increase at a uniform rate, but this is a useful approximate guide to the cost of living.

As inflation is not constant, governments often calculate tables to show the purchasing power of today's dollar in earlier years. In the United States, the Consumer Price Index (CPI) is calculated each month by finding the cost of a standard set of items (food, housing, vehicles, and so on). There are in fact several CPIs constructed. We shall always refer to the CPI-U, an index that reflects the cost of living in urban areas (about 80% of America). There is also a CPI-W, for wage-earners, and there are other indices. Tables of the CPI-U are available online at

http://stats.bls.gov/cpi/#tables.

The total CPI is divided by the average for 1982–1984 (the *base period*), and multiplied by 100. For example, the CPI for February 2006 was 198.7, so a collection of goods that cost $198.70 in February 2006 would have cost an average of about $100 in the base period. The average for a year is also published; the average for 1988 was 118.3; the figures for May and June were 118.0 and 118.5 respectively.

This can be used to compare two different years. The CPI for June 2002 was 179.9. The ratio $\frac{179.9}{118.5} = 1.518\ldots$ provides a comparison between June 2002 and June 1988 prices: if something cost $1,000 in 1988, our best guess is that it would cost about $1,518 in 2002. These figures are approximate, because the prices of different items do not increase at the same rate. However, it is reasonable to say "the cost of living was about 50% higher in 2002 than in 1988." A speaker in 2002 might say, "A dollar today is worth about two-thirds of what it was worth in 1988."

Suppose the cost of a major item at time $A$ is $\$X_A$. Suppose the CPI at time $A$ is $C_A$, and at time $B$ it is $C_B$. Then your estimate of the cost at time $B$ is

$$\frac{C_B}{C_A}X_A.$$

**Sample Problem 16.3** *A house cost $150,000 in June 1988. What would you expect a similar house to have cost in February 2006?*

**Solution.** If the house cost $150,000 in mid-1988, it is equivalent to a house that cost $\frac{198.7}{118.5} \times \$150,000 = \$251,518.99$ in February 2006. Your realistic answer might be "about $250,000."

**Your Turn.** A house costs $225,000 in February 2006. What would a similar house have cost in June 2002?

## 16.3  Animal Populations

Another example of exponential growth is the growth of an animal population. Given two animals (male and female), we know how frequently they will reproduce on average, and how many offspring will be produced. These numbers are not precise, but with large numbers the errors average out. If the animals reproduce an average of three offspring per year, and on average two die per year, the end result is as if the number of animals grows by 50% annually.

Of course, the animals do not all reproduce at the same time. The process is more like continuous compounding. In the example, the appropriate model is continuous compounding with an APY of 50%.

This model is more accurate with shorter breeding periods. When studying microscopic creatures, that reproduce within hours, reasonable predictions can be made of the population growth over periods of shorter than a day. For insects, a few days is often long enough for an accurate model. With humans, we need decades or even centuries. The "continuous compounding" model of a human population is used only for predicting the population movement in large cities, states or whole countries, because population fluctuations, caused by economic factors, the availability of highways, and so on, interfere with the model.

**Sample Problem 16.4** *A fish population doubles every year. At present it is 10,000. Approximately when will it reach 100,000? When will it reach 1,000,000?*

**Solution.** After $n$ years, the total population is $10,000 \times 2^n$, so the questions are, *"when is $2^n = 10$?"* and *"when is $2^n = 100$?"*

Now $2^3 = 8, 2^4 = 16, 2^6 = 64, 2^7 = 128$, so the answers are
$$100,000: \text{during the 4th year;}$$
$$1,000,000: \text{during the 7th year.}$$

## 16.4  Radioactive Decay

Radioactive decay works like continuous compounding in reverse. A radioactive material will dissipate with time, its molecules breaking down into molecules of other substances. If you have a certain amount present at a given time, it is found that the proportion that is lost depends only on the type of material and the time elapsed.

If you start with 100 g, then at the end of 1 day there will remain $100k$ g, where $k$ is a constant, between 0 and 1, depending only on the material. After $n$ days, the amount remaining is $100k^n$ g. This is exactly the same formula as continuous compounding, with $e^r = k$.

Of course, if $e^r$ is to be smaller than 1, $r$ must be negative. So radioactive decay is an example of exponential growth with a negative exponent.

The *half-life* of an element is the time it takes for the amount of it present to halve. For example, if you have 500 g with half-life 1 year, there will be 250 g after 1 year, 125 after 2 years, and so on. After $n$ half-lives amount $A$ decays to $A/2^n$.

**Sample Problem 16.5**  *An artificial element has a half-life of 1 h. You have 450 g. Approximately how long will it take until only 50 g is left?*
**Solution.** You want $450/2^n = 50$.
$$n = 3 : 450/2^3 = 450/8 = 56.25,$$
$$n = 4 : 450/2^4 = 450/16 = 28.125,$$
so the approximate answer is: a little over 3 h.
**Your Turn.** An artificial element has a half-life of 1 day. You start with 240 g. How much will be left after 4 days?

## Multiple Choice Questions 16

1.  Say your money grows from \$20,000 to \$22,000 over 2 years with continuous compounding. How much will you have after a further 2 years?

    A.  \$22,000        B.  \$24,000        C.  \$23,231        D.  \$24,200

2.  At noon Monday there are 10,000,000 bacteria present in a colony. At noon on Wednesday there are 12,100,000 present. How many do you expect to find at noon on Thursday?

    A.  11,000,000                    B.  24,200,000
    C.  14,641,000                    D.  13,310,000

3.  The population of a town was 150,000 at the beginning of 2012. Assuming an average growth rate of 2% per year, what is the expected population at the beginning of 2015?

    A.  150,000        B.  159,000        C.  159,181        D.  162,114

4.  The population of a small town was 1500 at the beginning of 2012. Assuming an average growth rate of 2.3% per year, what will the population be at the beginning of 2020?

    A.  1621        B.  1799        C.  2137        D.  3197

5.  The population of the state of Texas in 1993 was about 17,778,000. In 2001, the population was about 21,325,000. What was the average growth rate over that period of time?

    A.  2.3%        B.  1.2%        C.  1.9%        D.  0.3%

6.  The population of the US was 266 million at the beginning of 1998. Assuming an average growth rate of 0.6% per year, what would you expect was the population of the US at the end of 2001?

    A.  268.38 million                B.  270.82 million
    C.  272.44 million                D.  274.08 million

7.  An artificial element has a half-life of 3 h. If 640 g are present at noon, how much is left at 9 PM that day?

    A.  40 g        B.  80 g        C.  160 g        D.  480 g

8.  You have 400 g of a radioactive material in a container at noon on June 1. At noon on June 11, 100 g remain. How many grams were in the container at noon on June 6?

    A.  150        B.  200        C.  250        D.  300

# Exercises 16

1. $1,000 is invested at 10% annual interest, compounded continuously. What is the value of the investment after:

    (i)  three months;              (ii)  two years;              (iii)  ten years?

2. You invest $10,000 with continuous compounding. After 2 years, your investment is worth $11,200. What is its value after

    (i)  four years;                (ii)  ten years?

3. A house was bought for $120,000 in March 1991, when the CPI was 135.0. What is its expected value in February 2006, when the CPI was 198.7?

4. You invest $20,000 with continuous compounding. After 2 years, your investment is worth $23,000. What is its value after:

    (i)  four years;                (ii)  six years?

5. In January 2000 you have the choice of investing in houses for 5 years, or investing in a 5-year certificate of deposit that pays 3% compounded annually. The CPI for January 2000 was 168.8, and for January 2005 it was 190.7. Assuming the housing market shows the same growth as the CPI, which is the better investment? Why?

6. You bought your house for $167,000 in January 1999, when the CPI was 164.3. In December, 2005, when the CPI was 196.4, you were offered $195,000. Is this a good deal? Why?

7. The 2000 census shows the population of Cook County, IL as 5,376,741. If the population grows at the rate of 1.2% per year, what is the expected population in:

    (i)  2010;                      (ii)  2050?

8. A fish hatchery has 2,550 fish at the beginning of 2014. Each year the population grows by 50%. at the end of the year, 30% of the fish are sold. How many fish are there at the beginning of 2015? How many at the beginning of 2018 ?

9. The population of the state of Virginia was 7,187,700 at the beginning of 2002. Assuming an average growth rate of 1.5% per year, what will the population be at the beginning of 2020?

10. A certain insect doubles in population every week. There are 3,400 in a colony on Monday March 1st, and the numbers are checked every Monday. When do the numbers exceed 100,000?

11. A colony of penguins had 352 members at the beginning of 1998. The colony is expanding at an average rate of 1.2% per year. How many penguins will there be at the beginning of 2005?

12. You have 800 g of a radioactive element. Forty-two hours later there is 200 g remaining. What is the half-life of the material?

13. There are 1,280 g of an isotope present at noon on Monday. If the half-life is 12 h, how much is left at noon on Thursday?

14. An artificial element has a half-life of 1 h. You have 450 g. Approximately how long will it take until only 50 g is left?

# Your Turn Solutions

1.1  Three possibilities are $\{x : x^3 - 6x^2 + 11x - 6 = 0\}$, $\{y : y$ is positive, and $y^2 = 1, 4$ or $9\}$ and "the set of the first three positive integers." There are others.

1.2  $(1100100)_2 = 1 \times 2^6 + 1 \times 2^5 + 0 \times 2^4 + 0 \times 2^3 + 1 \times 2^2 + 0 \times 2^1 + 0$
$$= 64 + 32 + 0 + 0 + 4 + 0 + 0 = 100$$

1.3  $(10.111)_2 = 1 \times 2^1 + 1 \times 2^0 + 1 \times 2^{-1} + 1 \times 2^{-2} + 1 \times 2^{-3}$
$$= 1 \times 2 + 0 \times 1 + 1 \times .5 + 1 \times .25 + 1 \times .125$$
$$= 2 + .5 + .25 + .125 = 2.875$$

1.4
$$91/2 = 45, \text{ remainder } 1$$
$$45/2 = 22, \text{ remainder } 1$$
$$22/2 = 11, \text{ remainder } 0$$
$$11/2 = \phantom{0}5, \text{ remainder } 1$$
$$5/2 = \phantom{0}2, \text{ remainder } 1$$
$$2/2 = \phantom{0}1, \text{ remainder } 0$$
So $91 = (1011011)_2$.

1.5  $\sum_{i=3}^{5} i(i-1) = 3 \cdot 2 + 4 \cdot 3 + 5 \cdot 4 = 6 + 12 + 20 = 38;$
$\sum_{i=2}^{6} i = 2 + 3 + 4 + 5 + 6 = 20.$

1.6  (i) $1 + 3 + 5 + 7 + 9 = \sum_{i=1}^{5} 2i - 1;$    (ii) $8 + 27 + 64 + 125 = \sum_{i=2}^{5} i^3.$

1.9  The figures are represented by the following diagram. As only readers were surveyed, there is no need for any "outside area" outside $M \cup E$.

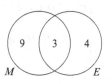

From the diagram we see that 16 readers were surveyed.

W.D. Wallis, *Mathematics in the Real World*, DOI 10.1007/978-1-4614-8529-2,
© Springer Science+Business Media New York 2013

1.11

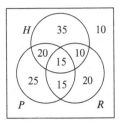

Twenty like horror and police procedural movies, but do not like romances. Twenty like romances only. Ten like none of these three types.

1.12    Write S for sedans and P for pickups.

|       | S  | P  |
|-------|----|----|
| Apr   | 32 | 16 |
| May   | 44 | 12 |

2.2    Equation (2.1) yields

$$|S \cup T| = |S| + |T| - |S \cap T| = 40 + 30 - 20 = 50.$$

From (2.2) we get

$$|S \setminus T| = |S| - |S \cap T| = 42 - 22 = 20.$$

2.3    Each digit can be chosen in 10 ways. So there are $10^4$ possibilities.

2.5    $6! = 6 \times 5 \times 4 \times 3 \times 2 \times 1 = 720.$

2.6    There are $9 \times 8 \times 7 = 504$ committees.

2.7    The boys can be ordered in $5! = 120$ ways. The girls can be ordered in $4! = 24$ ways. So there are $120 \times 24 = 2880$ arrangements.

2.8    $P(6,4) = 6 \times 5 \times 4 \times 3 = 360.$

2.9    $P(12,3) = 12 \times 11 \times 10 = 1320.$

2.10    The boys can be ordered in $3! = 6$ ways, and the girls can be ordered in $4! = 24$ ways. As the table is circular, it doesn't matter whether the boys are to the left or to the right of the girls. So there are $3! \times 4! = 6 \times 24 = 144$ arrangements.

2.13    There are three $A$'s, two $N$'s and one $B$, for a total of six letters. So the number of orderings is $6!/(3! \times 2!) = 60.$

2.14    $C(9,5) = \dfrac{9 \times 8 \times 7 \times 6}{4 \times 3 \times 2 \times 1} = 126, \binom{6}{0} = \dfrac{6!}{6! \times 0!} = 1.$

2.15    She must choose 4 of the last 7 questions, so $\binom{7}{4} = 35$ ways.

2.16    There are $C(8,4)$ choices of where to place the 1s, so the answer is $C(8,4)$, or 70.

2.17   You can choose the mysteries in $\binom{5}{2}$ ways and the westerns in $\binom{7}{3}$ ways. So you can choose in $\binom{5}{2} \times \binom{7}{3} = 10 \times 35 = 350$ ways.

2.18   The three consonants can be chosen in $\binom{5}{3} = 10$ ways, and the vowels in $\binom{3}{2} = 3$ ways. After the choice is made, the letters can be arranged in $5! = 120$ ways. So there are $10 \times 3 \times 120 = 3600$ "words."

3.1    $P(1) + P(2) + P(3) + P(4) = .20 + .16 + .16 + .16 = .68$.

3.2    $\{3, 4, 5, 6\}$.

3.5    Write $H$ and $T$ for heads and tails on the quarter, and 0, 1, 2 for the number of heads on the pennies. The sample space is $\{H0, H1, H2, T0, T1, T2\}$. The experiment consists of two parts. In the first the outcomes are $H$ and $T$; in the second there are three outcomes, 1, 2, 3. (The two pennies are not flipped separately, so in the case of one head we don't need to worry about which penny got the head and which the tail. Only the number of heads was recorded.) The tree diagram is

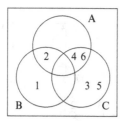

3.6

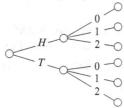

3.7   $A \cap C = \{HHT, HTH\}$, $B \cup C = \{HHH, HHT, HTH, HTT, THT, TTH\}$ and $\bar{A} = \{HHH, HTT, THT, TTH\} = B$

4.2    The scores, in ascending order, are

$$2, 3, 4, 4, 5, 6, 6, 6, 6, 7, 7, 7, 7, 7, 7, 9, 9, 10, 10, 10.$$

There are 20 scores. The median lies between the 10th and 11th score, each of which is 7. The median is $(7 + 7)/2 = 7$. The sum of the scores is 132, so the mean is $132/20 = 6.6$. The mode, or most common score, is 7.

4.3    The summary is $1, 4, 9, 12, 15$. The boxplot is

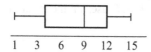

4.5

| Heads | 0 | 1 | | | 2 | | | 3 |
|-------|-----|-----|-----|-----|-----|-----|-----|-----|
| Rolls | TTT | HTT | THT | TTH | HHT | HTH | THH | HHH |
| Prob | .343 | .147 | .147 | .147 | .0963 | .063 | .063 | .027 |
| Sum | .343 | .441 | | | .189 | | | .027 |

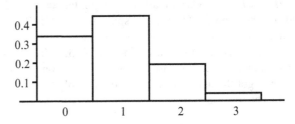

5.1   The mean sum is 10.5, and the standard deviation is 2.96. So you expect the mean of the sample to be 7, and the standard deviation of this mean is 0.296.

5.5   (i) 135 (27% of 500) (ii) $\sqrt{\frac{27\times73}{500}} = \sqrt{3.942} = 1.985$ (iii) $27 \pm 3.99\%$ (but $27 \pm 4\%$ or 23%–31% is close enough).

6.1

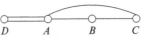

6.2   The walk must start (or finish) at $T$, because there is only one bridge to that island. With a little experimentation we find the solution $TXYZXYX$.

6.3

   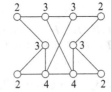

6.4   The original network is shown in figure (i). We start from $A$, and randomly choose the walk $AGJKHDA$. After these edges are deleted, figure (ii) remains.

We now start at $D$ (alphabetically, the first vertex remaining that was in the first walk and is not yet isolated). One walk is $DBCFEHGD$, and its deletion leaves figure (iii).

Finally, walk $FILKIHF$ uses up the remaining edges.

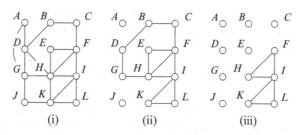

(i)                     (ii)                    (iii)

Putting these together, we get $AGJKHDBCFILKIHFEHGDA$.

6.5   The original is shown on the left. The two black vertices need another edge. They are not joined, so one edge will not suffice. So $eu(H) > 1$. The right-hand shows an Eulerization that requires just two edges. So $eu(H) = 2$, and the right-hand picture is a good Eulerization.

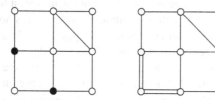

7.1   $(A,B)$,  $(A,B,C)$,  $(A,B,C,D)$,  $(A,B,C,D,E)$,  $(A,B,E)$,  $(A,B,E,D)$, $(A,B,E,D,C)$.

7.2   Candidates (i) and (iv) are Hamiltonian. Candidate (ii) is not Hamiltonian because it contains a repeated vertex, $b$. Candidate (iii) is not Hamiltonian because the graph contains no edge $cd$.

7.3   To traverse vertex $a$, a Hamiltonian cycle in $G$ must contain one of the paths $bad$, $bae$ or $ead$. If $bad$ is included, the other edge through $d$ might be $dc$ or $de$; in the former case, neither $bc$ nor $de$ can be edges, and the only cycle is $badcfe$, while the latter case bars $be$, and the only cycle is $badefc$. If $bae$ is included, $ad$ is not an edge, so $cd$ and $de$ are edges, so we have the path $baedc$, and the cycle is $baedcf$. If $ead$ is included, $de$ is not an edge, so $dc$ is an edge, and there are two possibilities, $adcbfe$ and $adcfbe$. So there are five Hamiltonian cycles.

Any Hamiltonian cycle in $H$ must contain edges $ab$ and $ad$, because $a$ has degree 2. This means $bd$ is not an edge (it would form a triangle), so $de$ must be in the cycle. There are two ways to finish a cycle: $bcfe$ or $bfce$. So there are two cycles: $dabcfe$ and $dabfce$.

7.4   The nearest-neighbor algorithm, starting from Evansville, begins with EM, because it has the least cost of the three edges incident with E. The next edge must have M as an endpoint, and ME is not allowed (one cannot return to E, it has already been used), so the cheaper of the remaining edges is chosen, namely MN. The only available edge from N is NS, as E and M have already been visited, and the route is EMNSE, with cost \$860.

Starting at Nashville, the first edge selected is NE, with cost \$180. The next choice is EM, then MS, then SN, and the resulting cycle NEMSN costs \$820.

If you start at St. Louis, the first stop will be Memphis ($220 is the cheapest flight from St. Louis), then Evansville, then Nashville, costing $820. From Memphis, the cheapest leg is to Evansville, then Nashville, and finally St. Louis, for $820. So both St. Louis and Memphis yield the same cycle as the Nashville case (with different starting points, and in reverse).

To apply the sorted edges algorithm, first sort the edges in order of increasing cost: EM($160), EN($180), NM($200), MS($220), ES ($240), NS($260). Edge EM is included, and so is EN. The next choice would be MN, but this is not allowed because its inclusion would complete a cycle of length 3 (too short), so the only other choices are MS and NS, forming route EMSNE (or ENSME) at a cost of $820.

In this example the route ENMSE, with cost $840, does not arise from either algorithm.

8.1   Graph (a) is not a tree because it is not connected; it has an isolated vertex. Graph (b) is a tree. Graph (c) contains a cycle, so it is not a tree.

8.2   If a tree has four vertices, then the largest possible degree is 3. Moreover there are three edges, so the sum of the four degrees is 6. As there are no vertices of degree 0 and at least two vertices of degree 2, the list of degrees must be one of $3,1,1,1$ or $2,2,1,1$. In the first case, the only solution is the star $K_{1,3}$. The only case with the second degree list is the path $P_4$. So there are two trees.

8.6   The first two choices are $ad$ and $be$, in either order. Third and fourth are $ab$ and $cf$, again in either order. $de$ would be next, but it is not allowed (it would form a cycle) so $ef$ is chosen; this completes the tree shown.

8.8

9.1   Call the missing digit $x$. Then $25 + x$ leaves remainder 5 when divided by 9. The only possibility with $0 \le x \le 9$ is $25 + x = 32$, so $x = 7$.

9.3   $5+9+4+0+1+3+3 = 25$; there are 2 large odd digits; $5+1+3+8+2+3 = 22$. So the sum is $(2 \times 25) + 2 + 22 = 74$; the check digit should be 6. The number is not legitimate.

9.4   In this case, the weighted sum comes to $254 = 11 \times 23 + 1$, so the check number is 10, written as $X$.

9.7   H500-0458-8827

9.8    H500-045-88-790-0

10.1    The final diagrams are

Encoded 1001                     Encoded 1110

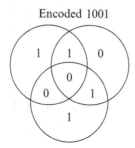                      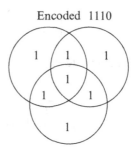

and the strings transmitted will be 1001101 and 1111111 respectively.

10.4    100 0 0100 01 1011 0 100

11.1    Use seven rows; you get CANYOUREADTHISMESSAGE, so you interpret it as CAN YOU READ THIS MESSAGE?

11.2    *CQQCAHKLU*

12.1    *A* receives 12 votes, *B* receives 11, *C* receives four. So (i) there is no majority winner (as there are 27 voters, 14 votes would be needed), and (ii) *A* is the plurality winner. In a primary election, *A* and *B* are selected to contest the runoff. For the runoff, *C* is deleted, so the preference profile is

| 7 | 5 | 8 | 3 | 4 |
|---|---|---|---|---|
| *A* | *A* | *B* | *B* | *B* |
| *B* | *B* | *A* | *A* | *A* |

,

or (combining columns with the same preference list)

| 12 | 15 |
|----|----|
| *A* | *B* |
| *B* | *A* |

.

So *B* wins the runoff.

12.2    (i) The votes for *A*, *B*, *C* and *D* are 13, 12, 7 and 11 respectively, so *A* would win under plurality voting.

(ii) Under the runoff method *A* and *B* are retained, and the new preference profile is:

| 6 | 7 | 7 | 7 | 2 | 7 | 5 | 2 |
|---|---|---|---|---|---|---|---|
| *A* | *B* | *A* | *A* | *A* | *B* | *B* | *B* |
| *B* | *A* | *B* | *B* | *B* | *A* | *A* | *A* |

,

giving 22 votes to *A* and 21 votes to *B*, so *A* wins again.

(iii) In the Hare method we first eliminate *C*, obtaining

| 6 | 7 | 7 | 7 | 2 | 7 | 5 | 2 |
|---|---|---|---|---|---|---|---|
| A | B | A | D | D | D | B | D |
| D | D | D | A | A | B | D | B |
| B | A | B | B | B | A | A | A |

.

Now we eliminate $B$:

| 6 | 7 | 7 | 7 | 2 | 7 | 5 | 2 |
|---|---|---|---|---|---|---|---|
| A | D | A | D | D | D | D | D |
| D | A | D | A | A | A | A | A |

.

So $D$ wins 30–13.

13.1  There are 32 voters, so the quota is 8. $A$ gets 10 first place votes, so $A$ is elected.

$A$ has a surplus of 2. The votes are divided in proportion 5:5 between $A$'s two lists, so we give one vote to each. After this allocation, the new table is

| 1 | 1 | 7 | 8 | 5 | 2 |
|---|---|---|---|---|---|
| B | E | B | C | D | E |
| E | B | D | D | C | B |
| D | D | E | E | E | C |
| C | C | C | B | B | D |

.

$B$ has eight votes, $C$ has eight, $D$ has five, and $E$ has three. No one meets the quota. So $E$ (who had the fewest votes) is eliminated. We now have

| 1 | 1 | 7 | 8 | 5 | 2 |
|---|---|---|---|---|---|
| B | B | B | C | D | B |
| D | D | D | D | C | C |
| C | C | C | B | B | D |

.

$B$ has 11 votes and is elected. The surplus of three votes is distributed in ratio $1:1:7:2$, approximately $0.27:0.27:1.91:0.55$. The table is

| 0.27 | 0.27 | 1.91 | 8 | 5 | 0.55 |
|------|------|------|---|---|------|
| D | D | D | C | D | C |
| C | C | C | D | C | D |

.

$C$ now has 8.55 votes and is elected. In total, $A$, $B$ and $C$ are elected.

13.4  Originally, $S$ is eliminated first, giving

| 8 | 6 | 5 |
|---|---|---|
| P | Q | R |
| Q | R | P |
| R | P | Q |

Next is $R$, leaving

| 8 | 6 | 5 |
|---|---|---|
| P | Q | P |
| Q | P | Q |

and $P$ wins.

If one voter switches, the profile becomes

| 8 | 5 | 1 | 5 |
|---|---|---|---|
| P | Q | R | R |
| Q | R | Q | S |
| R | P | P | P |
| S | S | S | Q |

Again $S$ is eliminated, leaving

| 8 | 5 | 1 | 5 |
|---|---|---|---|
| P | Q | R | R |
| Q | R | Q | P |
| R | P | P | Q |

But now $Q$ is eliminated, giving

| 8 | 5 | 1 | 5 |
|---|---|---|---|
| P | R | R | R |
| R | P | P | P |

and $R$ wins.

In a head-to-head competition between $Q$ and $R$, $Q$ would win 14–6 in the original election, and 13–7 using the modified preferences.

14.1    Use the formula $A = P(1 + nR/100)$ with $P = 1,200$ and $R = 10$. In the first case, $n$ is $\frac{1}{4}$, so

$$P = 1,200 \times (1 + .10/4) = 1,200 \times (1.025) = 1,230$$

and you repay $1,230. In the second case, $n$ is 3, so

$$P = 1,200 \times (1 + .10 \times 3) = 1,300 \times (1.3) = 1,560$$

and you repay $1,560.

14.2   From $A = P(1 + nR/100)$ with $A = 1,428$, $P = 1,400$ and $n = \frac{1}{3}$, we get

$$1,428 = 1,400 \times \left(1 + \frac{R}{300}\right) = 1,400 + \frac{14}{3} \times R,$$

and therefore $14 \times R = 28 \times 3, R = 6$. So the interest rate was 6%.

14.4   We have $P = 2,200$, $A = 2,530$ and $n = 2$. So

$$2,530 = 2,200 \left(1 + 2 \times \frac{R}{100}\right) = 2,200 + 44R; \ R = \frac{330}{44} = 7.5.$$

The rate was 7.5%.

14.6   For arithmetic growth,

$$1,200 \left(1 + \frac{10 \times 4}{100}\right) = 1,200 \times 1.4 = 1,680$$

so the interest is $480. Under geometric growth, the amount is

$$1,200 \left(1 + \frac{10}{100}\right)^4 = 1,200 \times 1.1^4 = 1,200 \times 1.4641 = 1,756.92$$

and the interest is $556.92.

14.8   In this case $P = 40,000$, $R = 12$, $t = 4$ and $N = 15$. So

$$P \left(1 + \frac{R}{t \times 100}\right)^{tN}$$

becomes

$$40,000 \times \left(1 + \frac{12}{400}\right)^{60} = 40,000 \times (1.03)^{60} = 0,000 \times 5.8916$$

which comes to $235,664.12\ldots$ and you owe $235,664.

14.9   The amount owing from a $100 loan at the end of one year is

$$A = 100\left(1 + \frac{12}{12 \times 100}\right)^{12}$$

$$= 100 \times 1.01^{12}$$

$$= 112.68,$$

so the APY is 12.68%.

15.1 Again, 6% annual interest is .5% per month, so $m = .005, D = 100$, but in this case $n = 12$, so you get

$$\$100(1.005^{10} - 1.005)/.005$$

$$= \$100(1.05114 - 1.005) \times 200$$

$$= \$20,000(0.04614)$$

$$= \$922.80$$

15.2 We use the formula

$$P(1+m)^n = \frac{D \times [(1+m)^n - 1]}{m}$$

with $P = 12,000$, $n = 60$ and $M = \frac{2}{3}, (1+m) = \frac{151}{150}$. Now

$$\left(\frac{151}{150}\right)^{60} = 1.4858457\ldots,$$

so

$$12,000 \times \left(\frac{151}{150}\right)^{60} = \frac{D \times [(\frac{151}{150})^{60} - 1]}{\frac{1}{150}}$$

becomes

$$12,000 \times 1.4858457 = 150 \times D \times .4858457;$$

that is,

$$D = 12,000 \times 1.4858457/(150 \times .4858457) = 244.676$$

and your payment is $244.68.

15.3 The interest rate is $1/150$ per month. So the accumulation after 24 years at 8% is

$$\$100,000 \times (151/150)^{288}$$

or $677,763.55. If you deposit $D per month, your accumulated savings would be

$$\frac{\$D \times [(151/150)^{288} - 1)]}{1/150} = \$D \times 5.77763554 \times 150 = \$D \times 866.645$$

So your monthly payment is $D = 782.05$.

15.4   $m = 7.5/12\% = 5/8\% = 1/160, D = 200, n = 60$. So

$$mA(1 + m)^n = D \times [(1 + m)n - 1]$$

becomes

$$A(161/160)^{60}/160 = 200[(161/160)^{60} - 1],$$

So $A \times 1.45329 = 160 \times 200 \times 0.45329$, $A = 9,980.99$, and you can afford about $9,981.

15.5   Simple interest is 3% per month, or 9% for the whole period. This comes to $54, so the total to be paid is $654. There are three monthly payments. Therefore, the monthly payment will be $654/3 or $218.

16.1   In this case $P = 40,000, R = 12, N = 15$ and $r = .12$. So

$$Pe^{rN} = 40,000 \times e^{1.8} = 241,985.89\ldots$$

and you owe $241,986. Compare this with the result of Your Turn Problem 14.8, where compounding was quarterly and the answer was $235,664.

16.2   Suppose the rate is $r$. Then $40,000e^{7r} = 80,000$ so $e^{10r} = 2$. Therefore $e^r$ equals the seventh root of 2, which equals 1.1041 approximately. The APR is approximately 10.4%.

16.3   $\frac{118.3}{198.7} \times \$225,000 = \$133,958.23$.

16.5   After 1 day, you have 120 g; after 2 days, 60; after 3 days, 30. So after 4 days there are 15 g remaining.

# Answers to Multiple Choice Questions

**Chapter 1**
1. C   2. B   3. D   4. C   5. A   6. C   7. A   8. C   9. B   10. A

**Chapter 2**
1. D   2. A   3. C   4. B   5. C   6. C   7. B   8. C   9. A   10. C

**Chapter 3**
1. C   2. B   3. D   4. B   5. B   6. A   7. D   8. A   9. A   10. A

**Chapter 4**
1. A   2. C   3. D   4. B   5. D   6. A   7. C   8. B   9. C   10. D   11. C   12. D
13. B

**Chapter 5**
1. D   2. A   3. B   4. B   5. C   6. A   7. C   8. D   9. C   10. A

**Chapter 6**
1. C   2. B   3. B   4. D   5. C   6. C   7. B   8. A   9. A   10. B   11. B   12. B

**Chapter 7**
1. B   2. A   3. D   4. A   5. C   6. B   7. C   8. A   9. C   10. B

**Chapter 8**
1. A   2. A   3. A   4. B   5. D   6. B   7. D   8. C   9. B   10. C   11. D   12. A

**Chapter 9**
1. A   2. A   3. B   4. A   5. A   6. A   7. B   8. B   9. A   10. D

**Chapter 10**
1. B   2. C   3. B   4. C   5. A   6. D   7. C   8. C   9. B   10. A   11. B   12. D

**Chapter 11**
1. B   2. C   3. B   4. C   5. D   6. B   7. D   8. B   9. D

**Chapter 12**
1. C   2. C   3. C   4. B   5. D   6. B   7. E   8. B   9. D   10. B

**Chapter 13**
1. A   2. A   3. C   4. D   5. C   6. B

W.D. Wallis, *Mathematics in the Real World*, DOI 10.1007/978-1-4614-8529-2,
© Springer Science+Business Media New York 2013

**Chapter 14**
1. C   2. B   3. C   4. D   5. A   6. B   7. C   8. B   9. D
**Chapter 15**
1. C   2. A   3. B   4. A   5. A   6. B   7. C   8. D   9. B   10. D
**Chapter 16**
1. D   2. D   3. C   4. B   5. A   6. C   7. B   8. B

# Answers to Exercises

## Chapter 1

1. (i) $2,4,6,8,10$ (ii) $-5,-3,-1,1,3,5$ (iii) Red, White, Blue (iv) January, June, July (v) M, I, S, P
2. Examples include "solution set of $x^3 + 3x^2 + 2x = 0$," "$\{0,1,2\}$," "$\{x : x \in \mathbb{Z}, 0 \le x \le 2\}$"
3. (i) and (iv)
4. (i) and (iii)
5. (i) $\mathbb{Q}, \mathbb{R}$ (ii) $\mathbb{R}$ (iii) $\mathbb{Z}, \mathbb{Q}, \mathbb{R}$ (iv) $\mathbb{N}, \mathbb{Z}, \mathbb{Q}, \mathbb{R}$ (v) none (vi) $\mathbb{Q}, \mathbb{R}$
6. (i) $\mathbb{Q}, \mathbb{R}$ (ii) $\mathbb{R}$ (iii) $\mathbb{N}, \mathbb{Z}, \mathbb{Q}, \mathbb{R}$ (iv) $\mathbb{R}$ (v) $\mathbb{Q}, \mathbb{R}$ (vi) $\mathbb{R}$ (vii) $\mathbb{Z}, \mathbb{Q}, \mathbb{R}$ (viii) $\mathbb{R}$
7. (i) 55 (ii) 34 (iii) 136 (iv) 0.25 (v) 136.25 (vi) 51.25
8. (i) 46 (ii) 44 (iii) 84 (iv) 0.625 (v) 44.625 (vi) 84.625
9. (i) 100011 (ii) 10011 (iii) 11001001 (vi) 10110101
10. (i) 10110 (ii) 101011 (iii) 100111000 (vi) 1011011
11. (i) $2+5+10+17+26+37 = 97$ (ii) $.1 + .01 + .001 = .111$
    (iii) $0+4+10+18+28+40+54 = 154$ (iv) $\frac{1}{2} + \frac{1}{3} + \frac{1}{4} + \frac{1}{5} = \frac{77}{60}$
    (v) $4 - 8 + 16 = 12$ (vi) $0+2+0+2 = 4$
    (vii) $1 + \frac{1}{2} + \frac{1}{3} + \frac{1}{4} = \frac{25}{12}$ (viii) $4+7+10+13+16+19 = 69$
12. (i) $1+4+9+16 = 30$ (ii) $0+2+0+2+0 = 4$ (iii) $2+3+4+5 = 14$
    (iv) $4+4+4+4+4+4+4+4+4+4 = 40$ (v) $6+12+20+30 = 68$
    (vi) $\frac{1}{4} + \frac{1}{9} + \frac{1}{16} = \frac{61}{144}$
13. (i) $\sum_{i=1}^{5} 3i - 2$ (ii) $\sum_{i=1}^{3} 4i - 2$ (iii) $\sum_{i=0}^{4} (-1)^i (3i+1)$ (iv) $\sum_{i=0}^{4} (-1)^i i^2$
    (v) $\sum_{j=1}^{6} 5 + 2(-1)^i$ (vi) $\sum_{i=1}^{4} i^2 - 1$ (There are other correct answers.)
14. (i) $\sum_{i=1}^{5} i^2$ (ii) $\sum_{i=0}^{3} (3i+2)$ (iii) $\sum_{i=1}^{4} (-1)^i i$ (iv) $\sum_{i=0}^{5} (6-i)$
    (v) $\sum_{i=0}^{4} 3^i$ (vi) $\sum_{i=0}^{5} (-1)^i (2i+4)$ (vii) $\sum_{i=0}^{4} (1+3^i)$ (viii) $\sum_{i=1}^{5} 4i$
    (ix) $\sum_{i=1}^{6} -i$ (x) $\sum_{i=1}^{3} 6i + 5$ (There are other correct answers.)
15. (i) $A_1 \subseteq A_4$, $A_1 \subseteq A_5$, $A_2 \subseteq A_5$, $A_4 \subseteq A_1$, $A_4 \subseteq A_5$, $A_1 = A_4$
    (ii) $A_1 \cap A_2 = \{2,4\}$, $A_1 \cap A_3 = \{3,4\}$, $A_1 \cap A_4 = \{1,2,3,4\}$,
    $A_1 \cap A_5 = \{1,2,3,4\}$, $A_2 \cap A_3 = \{4,8\}$, $A_2 \cap A_4 = \{2,4\}$,
    $A_2 \cap A_5 = \{2,4,6,8\}$, $A_3 \cap A_4 = \{3,4\}$, $A_3 \cap A_5 = \{3,4,5,8\}$,
    $A_4 \cap A_5 = \{1,2,3,4\}$

W.D. Wallis, *Mathematics in the Real World*, DOI 10.1007/978-1-4614-8529-2,
© Springer Science+Business Media New York 2013

16. (i) $A_1 \subseteq A_2$,   $A_1 \subseteq A_4$,   $A_4 \subseteq A_1$,   $A_4 \subseteq A_2$,   $A_1 = A_4$
    (ii) $A_1 \cap A_2 = \{1,2,4\}$,   $A_1 \cap A_3 = \{1,4\}$,   $A_1 \cap A_4 = \{1,2,4\}$,
    $A_2 \cap A_3 = \{1,3,4\}$,   $A_2 \cap A_4 = \{1,2,4\}$,   $A_3 \cap A_4 = \{1,4\}$

17. (i) $\{2,3,5\}$    (ii) $\{2,3,5,6,7,8,9\}$    (iii) $\{1,2,3,4,5,6,7,8,9\}$
    (iv) $\{2,3,5,6,8,9\}$

18. (i) $\{a,b,c,d,e,g,i\}$    (ii) $\{c,e\}$    (iii) $\{a,b,c,d,e,f,g,i,o\}$
    (iv) $\{a,b,c,d,e,i\}$

19. (i) True    (ii)   False    (iii) True    (iv) True

20. (i) True    (ii) True    (iii) True    (iv) False

21. (i) 675 (ii) 325

22. 20

23. 10

24. (i)                                                (ii) 3    (iii) 30

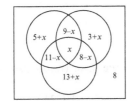

25. $M = $

|   | I | G |
|---|---|---|
| P | 4 | 24 |
| C | 6 | 12 |

26. $M = $

|    | A | B | C |
|----|-----|-----|-----|
| I  | 200 | 100 | 250 |
| II | 250 | 150 | 350 |

27. Examples:  $\sqrt{2}$,  $\pi$

## Chapter 2

1.  22; 11
2.  6; 16
3.  21; 51
4.  50; 32
5.  (i) 350 (ii) 475
6.  (i) 575 (ii) 425
7.  50; 150
8.  (i) 12 (ii) 14
9.  (i) $26^3 \times 10^3 = 17,576,000$ (ii) $24^3 \times 10^3 = 13,824,000$
10. 144
11. (i) $3^8 = 6,561$ (ii) $4^8 = 65,536$
12. $5 \times 4 \times 3 \times 2 \times 1 = 120$
13. (i) 336 (ii) 24 (iii) 120 (iv) 72
14. (i) $5! \times 3! = 720$ (ii) $4! \times 4! = 576$
15. (i) $15!$ (ii) $3! \times 4! \times 5! \times 6!$

16. (i) $P(9,4) = 3024$ (ii) 1008 (one-third of the possibilities) (iii) 1344
    (four-ninths of the possibilities) (iv) $P(7,4) = 840$
17. (i) $P(9,2) = 72$ (ii) $72/2 = 36$
18. (i) $7! = 5040$ (ii) $6! = 720$
19. (i) $7!/2! = 2520$ (ii) $7!/2!2! = 1260$ (iii) $6!/2!2! = 180$ (iv) $9!/3!3! = 9840$
    (v) $9!/4! = 15120$ (vi) $9!/2!3! = 30240$
20. $5 \times 5 \times 4 \times 4 \times 6! = 288,000$
21. $\{A,B,C\}, \{A,B,D\}, \{A,B,E\}, \{A,C,D\}, \{A,C,E\}, \{A,D,E\}, \{B,C,D\},$
    $\{B,C,E\}, \{B,D,E\}, \{C,D,E\}$
22. (i) 56 (ii) 126 (iii) 20 (iv) 35 (v) 1 (vi) 28
23. $C(6,2) \times C(12,3) = 3,300$
24. (i) $C(12,9) = 220$ (ii) $C(10,8) + C(10,8) + C(10,7) = 45 + 45 + 120 = 210$
    (or $C(12,9) - C(10,9) = 220 - 10 = 210$)
25. (i) $C(49,5)$ (ii) 44
26. 20
27. (i) $C(32,5) = 201,376$ (ii) $C(18,3) \times C(14,2) = 74,256$ (iii)
    $C(18,3) \times C(14,2) + C(18,2) \times C(14,3) = 129,948$
28. (i) $C(25,5) = 53,130$ (ii) $C(6,5) = 6$ (iii) $C(6,3) \times C(6,2) = 300$ (iv)
    $C(24,4) = 10626$ (v) $\sum_{i=0}^{3} C(6,4-i) \times C(6,1+i) = 780$
29. (i) $C(100,6) = 1,192,052,400$ (ii)
    $C(51,3) \times C(49,3) = 20,825 \times 18,424 = 383,679,800$

## Chapter 3

1. 0.22
2. 0.15
3. 1/32, or 0.3125
4. $HHH, HTH, HHT, THH, HTT, THT, TTH, TTT$
5. (i) 1/5 (ii) 4/5
6. (i) 1/6 (ii) 5/6 (iii) 1
7. (i) 5/36 (ii) 1/2 (iii) 1/4
8. (i) 5/16 (ii) 1/2 (iii) 1/2
9. 5/18
10. (i) $(2,0), (2,1), (2,2), (1,0), (1,1), (1,2), (0,0), (0,1), (0,2)$ (ii)
    $E = (2,0), (2,1), (1,0)$ $F = (2,0), (1,1), (0,2)$
    $G = (2,0), (2,2), (1,1), (0,0), (0,2)$
11. (iia) 0.25 (iib) 0.15 (iic) 0.35
12. (i) An initial vertex with three branches, terminating in vertices $R, Y, B$; and
    attached to each of these, a set of three branches, again terminating in vertices
    $R, Y, B$
    (ii) 9 (iii) $\{RR, YY, BB\}$

13.
(i) (i)

(ii) $\{H,TH,TT\}$ (iii) 1/4

14. (i) Yes (ii) No; add to 1.4, not 1 (iii) Yes (iv) No, contains a negative (v) no,
contains a "probability" greater than 1 (and adds to more than 1).

15. 0.55

16. (i) 36 (iia) $E = \{13,22,31\}, F = \{22,24,26,42,44,46,62,64,66\}$,
$G = \{31,32,33,34,35,36\}$ (iib) $F$ and $G$. (iii) The diagram consists of an
initial vertex with six branches, terminating in vertices $1,2,3,4,5,6$; and
attached to each of these, a set of six branches, again terminating in vertices
(iv)       11   12   14   15   16   21   23   25   41   43
           45   51   52   53   54   55   56   61   63   65

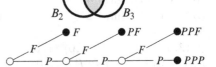

|   |   |   |   |   |   |
|---|---|---|---|---|---|
| 24 | 42 62 |   | 13 |   | 32 33 |
|   | 44 64 | 22 | 31 |   |   | 34 |
| 26 | 46 66 |   |   |   | 36 35 |

$1,2,3,4,5,6.$      $F$    $E$    $G$

(v) $P(E) = 1/12, P(F) = 1/4, P(G) = 1/6$

17. (i) 1/6 (ii) 5/9

18. (i) 1/2 (ii) 1/13 (iii) 1/26 (iv) 7/13

19. (i) 1/4 (ii) 1/4

20. (i) $\{BBB, BBG, BGB, BGG, GBB, GBG, GGB, GGG\}$
(ii) (a) $E = \{BBB, BBG, BGB, BGG\}$, $F = \{BBG, BGB, GBB\}$
(b) The oldest is a boy, the other two are one boy, one girl.
(iii) In the following diagram, $B1$ means "oldest is a boy," $B2$ means "middle
child is a boy," $B3$ means "youngest is a boy."

(i) Solid dots are outcomes:

21.

(ii) 4 (iii) $\{F, PF\}$

22. 2.81

23. 1.45

## Chapter 4

1. (i) mean $= 2.1$; median $= 1.5$; mode $= 5$; (ii) mean $= 3$; median $= 3$;
mode $= 4$

2. (i) mean $= 2.1$; median $= 1.5$; mode $= 1$; (ii) mean $= 4$; median $= 3$;
mode $= 7$

3. (i) mean $= 3$; median $= 3$; no mode; (ii) mean $= 4$; median $= 4$; mode $= 1$

4. (i) mean $= 4$; median $= 3$; mode 3; (ii) mean $= 3$; median $= 2.5$; no mode

5. 4

6. The histogram is roughly symmetric.

| 5 – 9 | 7, 8 |
|---|---|
| 10 – 14 | 11,11,11,12,12,12,14 |
| 15 – 19 | 17,18,18,18,18,19 |
| 20 – 24 | 20,20,21,22,22,22,23,23,23,24 |
| 25 – 29 | 25,25,26,26,26,26,27,27,28,28,28,29 |
| 30 – 34 | 30,30,31,32,32,32 |
| 35 – 39 | 35,36,36,38,38 |
| 40 – 44 | 40,42 |

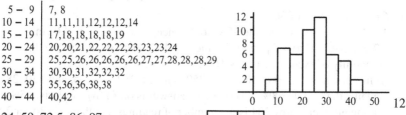

7. 24, 59, 72.5, 86, 97

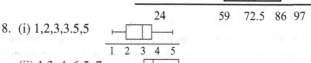

24     59   72.5  86  97

8. (i) 1,2,3,3.5,5

(ii) 1,3, 4, 6.5, 7

9. (i) 11, 1, 2.5, 4, 5

(ii) 7, 11, 12, 14, 17

10. (i) 11, 13, 16, 21, 28

(ii) 1,1.5, 4, 6.5, 9

11. $P(0) = 1/16 \ (TTTT)$
    $P(1) = 4/16 = 1/4$
    $(TTTH,TTHT,THTT,HTTT))$
    $P(2) = 6/16 = 3/8$
    $(TTHH,THTH,THHT,HTTH,HTHT,HHTT))$

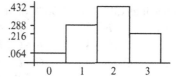

$P(3) = 4/16 = 1/4 \ (THHH,HTHH,HHTH,HHHT))$
$P(4) = 4/16 = 1/4 \ (HHHH))$

12. $P(0) = (2/5)^3 = .064$
    $P(1) = (2/5)^2(3.5) = .288$
    $P(2) = (2/5)(3.5)^2 = .432$
    $P(3) = (3.5)^3 = .216$

13. 4.5 years; 3.25–4.75 years
14. Mean $+\frac{2}{3}$SD $= 16+2 = 18$ h.
15. (i) The businesswoman is confusing the median with the mean. All we know from the information is that the middle-priced trip costs $600. Others could have been much less or much more.
    (ii) The top executives could have very large salaries. When they are averaged with other workers, 90% of the salaries could easily be below the mean.

## Chapter 5

1. 6 days (6 is 20% of 30).
2. $4 \pm 2 \times .01 = 4 \pm .02$, so the 95% confidence interval is 3.98–4.02 in.
3. $3.7 = 3.3 + 0.4$, so 3.7 pounds is "mean + 1 standard deviation," and the probability of exceeding this is 16%.
4. (i) In days, $\mu = 266$, $\sigma = 16$. So (i) $250 = \mu - \sigma$, $282 = \mu + \sigma$, and 250–282 contains the middle 68% of scores; the answer is 68%; (ii) similarly, 95%; (iii) the number below 234 is the number of readings less than $\mu - 2\sigma$, or 2.5%.
5. (i) Mean 16, sum of squared deviations 74, variance $74/12$,
    $SD = 2.483\ldots = 2.5$ approx.
    (ii) Mean 16, SD $\sqrt{(74/11)} = 2.593\ldots = 2.6$ approx.
6. (i) All mayonnaise jars available from the supplier. (ii) The hundred jars tested.
7. (i) The population is all Americans who live in a large city and are old enough to be classified as adults. (ii) The sample consists of those among the 500 interviewed who qualify as members of this population; children are not included, even if they smoke.
8. 800
9. The 95% confidence interval is 82% $\pm$ 6%, so the mean is 82% and the margin of error is 6%.
10. $44,000–$52,000
11. 65%
12. (a) 16% (b) 20.6–24.0 in.
13. (i) 1/10 or 10%. (ii) In simple random sampling, every sample has an equal chance of being chosen. In this example, some samples—for example, ones consisting of ten women—have zero probability of being selected.
14. Sample variability. The sample mean was less than two standard deviations above the population mean.
15. If there are $n$ interviewees, and the answer is 65% as expected, the standard deviation is $\sqrt{(65(100-65)/n)}$. Setting this equal to 2.5 and solving, the answer is 364 to the nearest integer.
16. $657/900 = 73\%$. The standard deviation is $\sqrt{\frac{73 \times 27}{900}} = \sqrt{2.19} = 1.48\%$. So the confidence interval is $73 \pm 2.96\%$, or 70.0%–76.0% to one decimal place.
17. Examples:

| A | B | C |   | A | B | C | D | E | F |
|---|---|---|---|---|---|---|---|---|---|
| B | C | A |   | B | C | D | E | F | A |
| C | B | A |   | C | D | E | F | A | B |
|   |   |   |   | D | E | F | A | B | C |
|   |   |   |   | E | F | A | B | C | D |
|   |   |   |   | F | A | B | C | D | E |

18. The survey will include non-students; students who do not drive are not represented. Also, those who frequent other parts of campus may have different opinions.
19. Normal-type examples include heights and weights of children in a class, incomes, number of insects per square foot, and so on. One type of

non-normal example is a case where two rather different groups are mixed. For example, if you have two brothers aged 6 and 14, the set of heights of all your brothers' classmates will not be approximately normal.

## Chapter 6

1. (i) 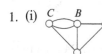    (ii)     (iii)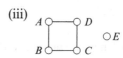

2. (i)     (ii)     (iii)

3. (i)     (ii)

4. (i) 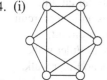    (ii)

5. (i)    (ii)

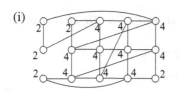

6. (i)    (ii)

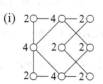   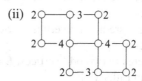

7. (i)    (ii)

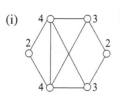

8.

(i)

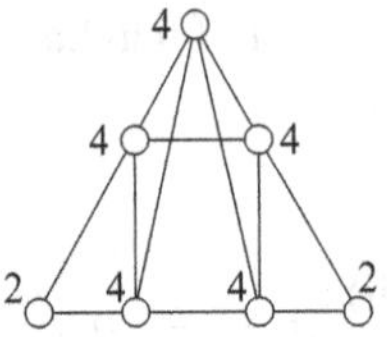


(ii)

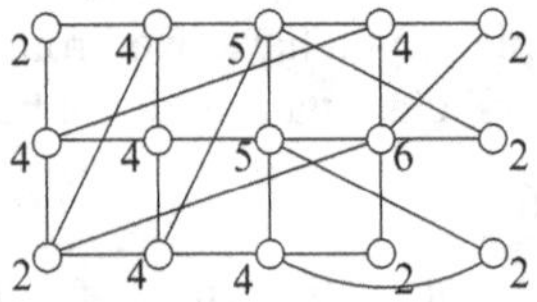


9. (i) There are no odd vertices; *ABGEFACBEGJKHFDCEJHDA* is an Euler circuit.
   (ii) There are two odd vertices, *C* and *H*. *HMLEMGBAEJKFEBCDGHC* is an Euler walk; there is no Euler circuit.
10. (i) There are four odd vertices, *A*, *B*, *D* and *E*. There is no Euler walk.
   (ii) There are no odd vertices. There is no Euler walk because the graph is not connected.
11. (i) There are two odd vertices, *B* and *L*. *BEHLKJFABCDGML* is an Euler walk; there is no Euler circuit.
   (ii) There are six odd vertices, *A*, *B*, *C*, *E*, *F*, *G*. So there is no Euler walk.
12. (i) There are no odd vertices; *ACFGEBCGDA* is an Euler circuit.
   (ii) There are four odd vertices, *A*, *C*, *E*, *F*. So there is no Euler walk.
13. (i) There are no odd vertices; *ABDFECBDECA* is an Euler circuit.
   (ii) There are no odd vertices; *DFCGLEABFKJEHD* is an Euler circuit.
14. (i) There are no odd vertices; *ACFEDBCEBA* is an Euler circuit.
   (ii) There are four odd vertices, *L*, *S*, *M* and *T*. There is no Euler walk.
15. (i) There are four odd vertices, *B*, *E*, *M* and *Q*. There is no Euler walk.
   (ii) There are eight odd vertices,*B*, *E*, *F*, *G*, *H*, *J*, *M*, *P*. So there is no Euler walk.
16. (i) There are no odd vertices; *ABCDHMGDMLKJEAFJA* is an Euler circuit.
   (ii) There are two odd vertices, *E* and *H*. *EABCDHMLKJEFADGMJFGH* is an Euler walk.
17. (i) There are two odd vertices. There is no Euler walk because the graph is not connected.
   (ii) There are two odd vertices, *E* and *H*. *EABFGHCDEJKH* is an Euler walk.
18. In each case, heavy lines indicate the edges to be duplicated.

(i)

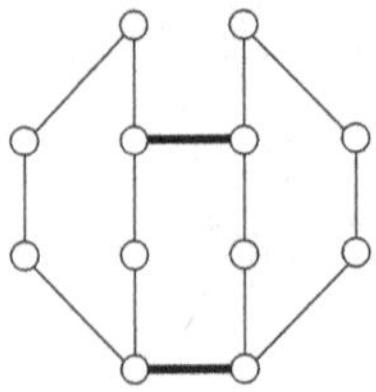

(ii)

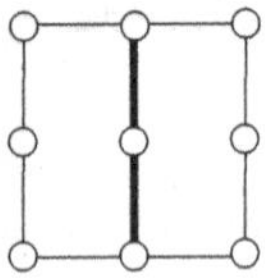

19. In each case, heavy lines indicate the edges to be duplicated.

(i) 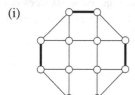          (ii)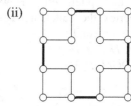

20. In each case, heavy lines indicate the edges to be duplicated.

(i) 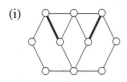          (ii)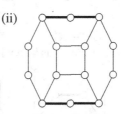

21. In each case, heavy lines indicate the edges to be duplicated.

(i) 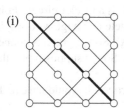          (ii)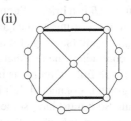

22. In each case, heavy lines indicate the edges to be duplicated.

(i)           (ii)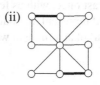

23. In each case, heavy lines indicate the edges to be duplicated.

(i)           (ii)

24. In each case, heavy lines indicate the edges to be duplicated.

(i) 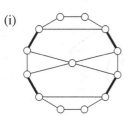          (ii)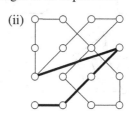

25. Point out to your friend that, in order to go from edge 1 to edge 2, the walk must start at the top left vertex. However, the last edge, edge 10, does not touch that vertex: the walk is not a circuit. Edge 10 touches edge 1, but it touches the wrong end.

26. *CDH, CGH, JKP, JNP.*

27. *AEJ, AFJ, DGM, DHM.*

28.

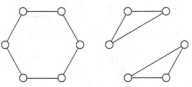

29.   (i)  Bridges are shown as heavy lines.

(ii) Assume *AB* is a bridge. Suppose the graph has an Euler circuit. Follow the circuit, starting with edge *AB*. If you delete that edge, you still have an Euler walk (starting at B and ending at A). So the graph is connected. This contradicts the fact that *AB* is a bridge. So the supposition—that there is an Euler circuit—is impossible.

(iii) The graph is connected. If it had no odd vertices, it would have an Euler circuit. So it must have some odd vertices.

30. If you construct a graph where the vertices represent airports serviced by the company and edges represent routes, then the representative wants to traverse all the edges at least once, with as few repetitions as possible. An efficient Eulerization could represent his route.

31. They would like to plan a route in which plows do not cover any roads twice, if possible.

## Chapter 7

1. (i) 6 (ii) 4
2. (i) 10 (ii) 6
3. 20160
4.

 (i) 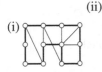   (ii)

5.   (i)                     (ii)

6.

(i)     (ii)

*abecfdgh, adbcfgeh, adbecfgh, adfcbegh, adgfcbeh*

7. In order to include the three vertices of degree 2, all the edges touching them must also be in the cycle. So the central vertex will have degree 3, which is impossible in a cycle.

8. In each case, in order to include the two vertices of degree 1, it is necessary to include the edges touching them. After this, it is easy to check that no Hamilton path is possible.

9. (i) *abcdef, abdcef* (ii) *abcehgfd, abcfdgeh, abdfcegh, abdgfceh,*

10. (i) *abdegfc, abedgfc, abegdfc, abegfdc, abegfcd, acfgebd* (ii) none

11. Cheapest is 24. Each NN is 27.

12. NN: *a* 126, *b* 126, *c* 129, *d* 130, *e* 129, SE: 129

13. NN: *a* 286, *b* 286, *c* 286, *d* 286, *e* 286, SE: 286

14. NN: *a* 158, *b* 151, *c* 155, *d* 151, *e* 151, SE: 151

15. NN: *a* 377, *b* 371, *c* 371, *d* 377, *e* 371, SE: 371

16. NN: *a* 158, *b* 152, *c* 158, *d* 158, *e* 152, SE: 158

17. NN: *a* 74, *b* 80, *c* 81, *d* 74, *e* 75, SE: 80

18. NN: *a* 125, *b* 137, *c* 152, *d* 149, *e* 125, SE: 137

19. NN: *a* 242, *b* 242, *c* 242, *d* 242, *e* 242, SE: 242

20. One example: take a cycle of length 8, say (*abcdefgh*), and add edges *be* and *df*.

21. Example: Suppose a car dealer in Chicago wants to deliver a car to New York, visiting several branches in towns along the way. The dealership supplies her with an air ticket from New York back to Chicago. A Hamilton path on a map with all the relevant dealerships marked would be an appropriate trip guide.

## Chapter 8

1. 7

2. (i) Say vertices *x* and *y* have degree 3. If they are not adjacent, each must be adjacent to three of the remaining vertices. Since the others have degree 1, none can be adjacent to both *x* and *y*, so there are at least six vertices of degree 1. This is impossible, so *x* and *y* are adjacent.

(ii) No. Even if the vertices of degree 4 are adjacent, the above argument shows there would have to be more than 4 vertices of degree 1.

3. There are two possibilities:

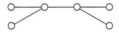    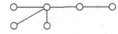

4. The degrees add to 12, and exactly three equal 1, The other degrees must be 2, 2, 2, 3. The solutions are

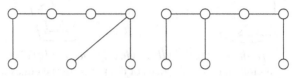

5. Here are typical solutions:

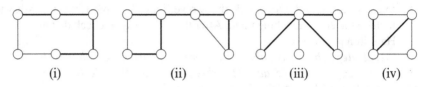

(i)                    (ii)                    (iii)                    (iv)

6. There are 22 trees.
7. There are 9 trees.
8. There are 15 trees.
9. There are 4 trees.

10. Total cost 25.
    Example:
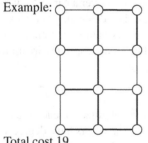

11. Total cost 15.
    Example:
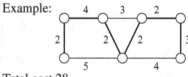

12. Total cost 39.
    Example:
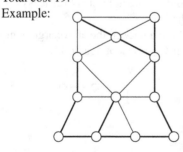

13. Total cost 28.
    Example:
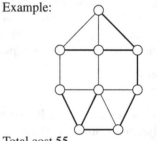

14. Total cost 19.
    Example:

15. Total cost 55.
    Example:
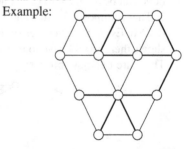

16. Total cost 37.
    Example: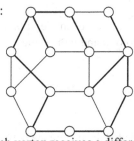

17. Total cost 45.
    Example: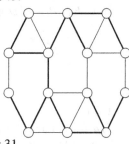

18. Total cost 45.
    Example: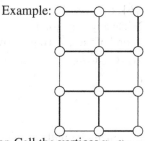

19. Total cost 31.
    Example:

20. In $K_5$, each vertex receives a different color. Call the vertices $x_1, x_2, \ldots, x_5$ and allot color $i$ to $x_i$. If edge $x_i x_j$ is deleted, recolor $x_j$ with color $i$.

21. A generalization of Exercise 1: in $K_5$, say vertex $x_i$ receives color $i$. Delete edge $x_i x_n$ is deleted, and recolor $x_n$ with color $i$.

22. 3 colors; example, classes $\{a,d\}, \{b\}, \{c\}$.

23. 6 colors; classes $\{a\}, \{b\}, \{c\}, \{d\}, \{e\}, \{f\}$.

24. 4 colors; classes $\{a,d\}, \{b,c\}, \{e\}, \{f\}$.

25. 3 colors; example, classes $\{a,d,g,j\}, \{b,c,f\}, \{e,h,i\}$.

26. 2 colors; classes $\{a,b,c\}, \{d,e,f\}$.

27. 5 colors; example, classes $\{a,f\}, \{b\}, \{c\}, \{d\}, \{e\}$.

28. $\chi(A) = 3, \chi(B) = 3, \chi(C) = 2$.

29. $\chi(D) = 3, \chi(E) = 2, \chi(F) = 3$.

30. 3 colors; color classes
    $\{A,C\}, \{B,D\}, \{E,F\}$
    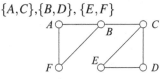

31. 3 colors; color classes
    $\{A,D\}, \{B,F\}, \{C,E\}$
    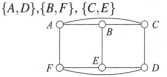

32. 2 colors; color classes
    $\{A,C,E,H\}, \{B,D,F,G\}$
    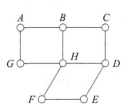

33. 4 colors; color classes
    $\{A,C,F\},\{B,E\},\{D,G\},\{H\}$

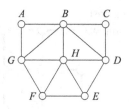

34. *A* or *C*.
35. (i) $a,b,d,e$ must all receive different colors, so the chromatic number is at
    least 4. One four-coloring uses classes $\{a,c\},\{b\},\{d\},\{e\}$; another uses
    $\{a\},\{b\},\{d\},\{c,e\}$.
    (ii) Delete any edge joining two of $a,b,d,e$. For example, if $ab$ is deleted, use
    the second of the colorings above but color $a$ and $b$ the same.
36. $a,e,b,f,c,g,d,h$.

## Chapter 9

1. 6
2. 1
3. Yes; the remainder will be changed provided the difference between the
   correct number and the error is not divisible by 9; divisibility by 9 is
   impossible when the correct digit is 3.
4. 7
5. 5
6. No; after the substitution the remainder on division by 9 will be unchanged.
7. 7
8. No; check digit should be 0.
9. 0
10. 0-1137-7251-3
11. No; check digit should be 1, not 3.
12. The new number 7-0167-2378-6 will yield a weighted sum 227, not a multiple
    of 11, so the system will detect an error, but it cannot recover the original: for
    example, if the first 6 is changed to 5, the resulting 7-0157-2378-6 is valid.
13. B424
14. D250
15. S512
16. W650
17. M625-5739-2696, 6255-7392-696M
18. P625-478-91-164-0
19. B451-9168-8844, B451-916-88-943-1
20. B600-2306-6125, B600-230-66-161-0

## Chapter 10

1. 1000110, 0100101
2. 1110100, 0101110.

3. 1011010, 1001101.
4. 1111111 1101000 1011010 0011010
5. 1011 (the whole string, after decoding, is 1011010).
6. 0001 1000 1010 ( the decoded whole strings are 0001011 1000110 1010001).
7. No, it returns 1010. Two errors were made.
8. 00010100
9. 4
10. Each is 5.
11. (i) 6 (ii) 5 (= 6−1) (iii) 2. (6−1)/2 = 2.5, so it can correct "at most 2.5 errors," but there is no such thing as half an error.
12. If the minimum distance is 2, the code can detect 2−1 = 1 error, and can correct at most 1/2; since you cannot correct half an error, it cannot correct errors.
13. (i) To see that the code is linear, it is necessary to check that the sum of any two words is again a word. This is so; the binary addition table is

|   | A | B | C | D | E | F | G | H |
|---|---|---|---|---|---|---|---|---|
| A | H | D | E | B | C | G | F | A |
| B | D | H | F | A | G | C | E | B |
| C | E | F | H | G | A | B | D | C |
| D | B | A | G | H | F | E | C | D |
| E | C | G | A | F | H | G | B | E |
| F | G | C | B | E | D | H | A | F |
| G | F | E | D | C | B | A | H | G |
| H | A | B | C | D | E | F | G | H |

(ii) 2 (iii) 2 (iv) The code can detect any one error but cannot correct an error; for example, if 1110 is received, it could have come from a single error in A, B, D or G.
14. (i) The binary addition table is the same as in the preceding question. (ii) 4 (iii) 4 (iv) it can detect 3 and correct 1.
15. (i) 11111111 and 00000000 (ii) 10011001, 11100001 and 00011110. (iii) 11111111 has weight 8, 00000000 has weight 0, and all other words have weight 4, so the code has weight 4. (iv) The minimum distance is 2 (for example, 00000000 and 10000001 are both codewords), so the code cannot correct errors. (v) 11111111 and 00110011 respectively.
16. 1010 111 11 0     0000 0 010 0     1 111 11 111 010 010 111 011
17. breakfast
18. 00101101110100110110111001010
19. AATCACTGAACA
20. ATCGATTCGACATCGACAA

## Chapter 11

1. *Encode* does not necessarily mean there is a secret key.
2. PLEASE SEND MORE ARMIES (circumference 5).
3. WAKE UP AND SMELL THE COFFEEE (circumference 4).
4. WHEN ROME WAS FIRST A CITY ITS RULERS WERE KINGS (circumference 3). (from Tacitus, "The annals of Imperial Rome")
5. HOW MANY MORE MILES MUST WE MARCH (circumference 3).
6. THE SPARTANS WERE THE INVENTORS OF THIS METHOD OF SECRET WRITING (circumference 6).
7. LIKE MANY OF THE GREAT GENERALS OF HISTORY, CAESAR SEEMS TO HAVE BEEN LACKING IN CRYPTOGRAPHIC SUBTLETY (circumference 7) (from Arnold Beutelspacher, "Cryptography")
8. VHQG WURRSV
9. WKH HQG LV QHDU
10. attack
11. retreat
12. AOL TVVU OHZ YPZLU
13. ZLCLU RUPNOAZ HYL HWWYVHJOPUN
14. do not pass go
15. bread and circuses
16. OFMCEPFQDKLSCEPWLPVALPOLVE
17. AKTFMDAOFPEKLQDAOVLORBLOAKTOYMQ
18. Modulo 5,
    (i) $3 + 5 \equiv 3$;          (ii) $4 \times 3 \equiv 2$;          (iii) $8 - 2 \equiv 1$.
19. 7.
20. 3, 7, 9 (1 is not technically wrong, but it is pointless)
21. 12
22. If we write A as "01", AA is 0101 and K is 11. If A is written "1," AA and K both come out as 11.

## Chapter 12

1. 18
2. 120
3. *A* wins under both methods.
4. *C* received 22 votes. There is no majority winner. *B* wins under plurality.
5. (i) 123 voters. (ii) Smith 52, Jones 36, Brown 35. (iii) Smith wins under plurality. (iv) There is no majority winner.
6. (i) 1193 voters. (ii) Andrews 224, Brown 442, Carter 527. (iii) There is no majority winner. (iv) Carter wins under runoff (Carter beats Brown 628–565).
7. (i) Jones wins under A, Brown wins under B. (ii) Brown wins under both methods.
8. *U* wins (*X* beats *Y* 20–10, *Z* beats *X* 18–12, *U* beats *Z* 22–8, *U* beats *V* 30–0).
9. (i) *S*. (ii) *C*. (iii) *C*.
10. (i) *B*. (ii) *C*.

11. (i) No winner. (ii) $C$. (iii) $A$.
12. $W$ and $Z$ are eliminated. The first-place votes are then $V$–10, $X$–6, $Y$–5. In the runoff, $X$ beats $V$ 11–10.
13. $C$ wins under plurality. $A$ wins a runoff, beating $C$ 9–7. $A$ is a Condorcet winner. runoff, $X$ beats $V$ 11–10.
14. (i) $B$. (ii) $B$. (iii) $A$. (iii) Yes, $B$ is a Condorcet winner.
15. (i) $A$. (ii) $B$. (iii) There is no Condorcet winner. $A$ wins under Condorcet's extended method.
16. Jones.
17. Brown.
18. Yes, $A$ is a Condorcet winner.
19. (i) $B$. (ii) $B$. (iii) $B$. (iv) $A$. (v) $B$. No Condorcet winner; $B$ wins under the extended method.
20. No Condorcet winner. (i) $A$. (ii) $C$. (iii) $B$. (iv) $B$.
21. There is no winner in the Hare method: $A$ and $B$ tie. $C$ is a Condorcet winner.
22. No difference.
23. $Z$ wins if $n = 3$. $X$ wins if $n = 5$, 6 or 7. $Y$ wins if $n \geq 8$. There is a tie when $n = 4$.
24. With four major parties, most seats will have four candidates, all of whom will have good support, so there will often be no majority winner.
25. Accurate percentages suffice.
26.

| 1 | 1 | 1 | 1 |
|---|---|---|---|
| $W$ | $X$ | $Y$ | $Z$ |
| $X$ | $Y$ | $Z$ | $W$ |
| $Y$ | $Z$ | $W$ | $X$ |
| $Z$ | $W$ | $X$ | $Y$ |

27. The Condorcet winner will beat every other candidate in a runoff. So that candidate will lead the preference list in the extended method.

## Chapter 13

1. Quota is 9. There are no winners initially. Then $D$ is eliminated. $D$'s five votes go to $C$. So $C$ is elected. All $C$'s surplus goes to $B$, and $B$ is elected.
2. Quota is 10. There are no winners initially. $D$ is eliminated. $C$ receives all five votes, for a total of 12, so $C$ is elected. The two surplus votes go to $B$, and $B$ is elected.
3. Quota is 10. $B$ is elected initially. There is one surplus vote each for $A$ and $D$. $C$ is eliminated, then $A$. Then $D$ is elected.
4. Quota is 13. $C$ is elected initially. 0.5 surplus votes go to $B$ and $D$ each. As no one else has reached the quota, $E$ is eliminated, and $B$ is elected.
5. Quota is 10. $A$ and $D$ are elected initially. $B$ receives more than one surplus vote, and is elected.
6. Quota is 25. No one is elected initially. so $B$ is eliminated and $A$ now has 40 votes and is elected. 9 of $A$'s votes go to $E$ and 6 to $D$; $E$ is now elected.
7. Quota is 8. $A$, $C$ and $E$ are elected.

8. Quota is 12. Both *A* and *C* are elected initially. After the surplus is allocated, *E* is elected.
9. (i) A tie between *B* and *F*. (ii) No one is chosen. (iii) A tie between *B* and *F*. (iv) Both *B* and *F* are chosen.
10. (i) *B*, *E* and *H* are chosen. (ii) Eight votes. (iii) *B* and *E* are chosen.
11. (i) No winner. (ii) *B*. (iii) *C*. (iv) Now *B* wins under the Borda count.
12. (i) *C*. (ii) *A*. (iii) *B*.
13. (i) *B* still wins (by a larger margin). (ii) There would now be a tie between *B* and *D*.
14. If they change their preference to *B*,*C*,*A* then *B* would win; they prefer this to *A* winning.
15. Initially *A* gets 57 points, *B* 69 and *C* 72, so *C* wins. If four swap *A* and *B*, *B* gets 73, and wins.
16. (i) *A* (ii) If they exchange *A* and *D*, *B* will win. (iii) If they exchange *D* and *C*, *C* and *A* will tie, and another election will be required; conceivably, *A* would not win. But this is not guaranteed.
17. (i) *A* (ii) *D* would then win.
18. (i) The runoff would be between *A* and *B*; *B* wins 9–8. (ii) After the change the runoff would be between *A* and *C*; *C* wins 10–7.
19. In any case the runoff would be between *Y* and *Z*; *Z* wins 26–20. (i) Initially *X* beats *Z* 24–22. (ii) The voters in the fourth column cannot change this.
20. In runoffs, *A* beats *B*, *B* beats *C* and *D*, *C* beats *A* and *D* and *D* beats *A* (margin is 2–1 in each case except *C* beats *D* 3–0). Some winning agendas (not unique): for *A*, *C*,*D*,*B*,*A*; for *B*, *C*,*A*,*B*,*D*; for *C*, *A*,*B*,*C*,*D* and for *D*, *C*,*B*,*A*,*D*.

## Chapter 14

1. (i) \$4,320 (ii) \$4,960 (iii) \$5,600
2. (i) \$480 (ii) \$720 (iii) \$1,440
3. (i) \$2,875 (ii) \$3,000 (iii) \$3,375
4. (i) \$300 (ii) \$600 (iii) \$750
5. (i) \$2,005 (ii) \$2,020 (iii) \$2,030
6. (i) \$20 (ii) \$60 (iii) \$120
7. 10%
8. 7.5%
9. 12%
10. 4%
11. (i) \$4,200 (ii) \$12,600 (iii) \$1,100
12. (i) \$3,200 (ii) \$2,800 (iii) \$6,600
13. 9 months
14. (i) \$1,125.51 (ii) \$1,123.60 (iii) \$1,120
15. (i) \$27.05 (ii) \$26.90 (iii) \$26.25
16. \$3,588.26
17. \$991.92
18. \$225.36

19. $2,338.10
20. $2,420
21. $7,441
22. 12 (actually about $11\frac{2}{3}$ years)
23. after 6 years
24. (i) 5.095% (ii) 6.136% (iii) 3.042%

## Chapter 15

1. $27,223.76
2. $24,664.64
3. $59,556.16
4. (i) $1,842.71 (ii) $1,665.58 (iii) $14,83.62
5. $1199.10
6. $91.68
7. (i) $2,501.52 (ii) $8,230.54 (iii) $15,084.83 (iv) $27,971.23
8. (i) $2,028.53 (ii) $1,609.25
9. $758.48
10. $200; $193.43; the credit union is better.
11. (i) 28,951.22 (about 29%) (ii) 45,996.46 (about 46%) (iii) 56,641.35 (about 57%)
12. $113.33
13. $236
14. (i) 40,913.65 (about 41%) (ii) 66,069.17 (about 66%)
15. $40,000; $666.67
16. $21,917.81; $608.83

## Chapter 16

1. (i) $1025.32 (ii) $1,221.40 (iii) $2,718.28
2. (i) $12,544 (ii) $17,623.42
3. $176622.22
4. (i) $26,450 (ii) $30,417.50
5. The CD (it returns $1,159.27 for each $1000 invested; housing returns $1129.74)
6. No, the value is approximately $199,627
7. (i) $5,376,741 \times (1.012)^{10}=6,057,930$ (ii) $5,376,741\times(1.012)^{50}=9,762,135$
8. $2,550 \times 1.2 = 3,060$; $2,550 \times 1.2^4 = 5287.68$ so answer is 5,287 or 5,288.
9. 9,396,772
10. April 5th
11. 382 or 383
12. 21 h: there are 800 g at the start, 800/2 = 400 after 21 h, 400 = 200/2 after 42 h.
13. 20 g
14. You want 450/2n = 50. If $n=3: 450/2^3 = 450/8 = 56.25$. If $n = 4: 450/2^4 = 450/16 = 28.125$. So the approximate answer is: a little over 3 h.

# Index

Printed in the United States
By Bookmasters